Hermetic Recreations

Hermetic Recreations

INCLUDING THE SCHOLIUM

TEXT ESTABLISHED, TRANSLATED,
AND ANNOTATED BY

Christer Böke
& John Koopmans

WITH CONTRIBUTIONS BY

Stanislas Klossowski de Rola
& Aaron Cheak

RUBEDO PRESS
AUCKLAND · MMXVIII

Hermetic Recreations: Including the Scholium
(Anonymous)

EDITED, TRANSLATED, AND ANNOTATED BY
Christer Böke, John Koopmans,
Stanislas Klossowski de Rola,
& Aaron Cheak

※

Published by

RUBEDO PRESS
PO BOX 21266
HENDERSON 0650
AUCKLAND · NEW ZEALAND

© Rubedo Press 2018

ISBN: 978-0-473-40989-0 (hardback)
ISBN: 978-0-473-41078-0 (paperback)

All rights reserved.
No part of this work may be reproduced
without express permission from the publisher.
Brief passages may be cited by way of criticism,
scholarship, or review, as long as full
acknowledgement is given.

※

Design and typography by Aaron Cheak
Cover illustration from the *Aurora Consurgens*,
Zürich Zentralbibliothek, Ms. Rh. 172,
fifteenth century (modified).

SCRIBE SANGUINE QUIA SANGUIS SPIRITUS

TABLE OF CONTENTS

Preface
7

Introduction
9

Transcription & Translation

Récréations Hermétiques	Hermetic Recreations
22	23
Scholies	Scholium
90	91

Bibliography
141

About the Contributors
149

Preface

THE PRESENT VOLUME CAME TOGETHER PRINCIPALLY through the dedicated efforts of Christer Böke and John Koopmans, who sourced the rare French manuscript of the *Récréations hermétiques*, provided a faithful transcription of it, and made an initial translation of the text into English. This initial translation was passed on to Stanislas Klossowski de Rola for verification. Not to do things by halves, Stanislas took it upon himself to provide a completely new translation, insisting that the text should preserve, rather than clarify, the many archaisms of the original. The manuscript, the transcription, and the two translations were then passed on to me, and my involvement in this project, in addition to being the publisher, has been to provide the final Hermetic synthesis. This was done by closely consulting, comparing, and translating the different texts. In most cases, I simply took the best that both translations had to offer in terms of accuracy and style, but I have also re-worked some particularly thorny passages—especially where both translations were at variance—and have polished the overall prose to reflect the dignity of the subject matter. The orthographic glosses to the French text were provided by John Koopmans, while the annotations to the English text reflect the contributions of all four translators.

AARON CHEAK, PHD

Introduction

THE TRANSLATED TEXT WHICH WE HAVE THE PLEASURE of introducing here for the very first time to an English audience is indeed a very curious and important piece of work, significantly more revealing in terms of explaining alchemical metaphors and secret processes than most alchemical texts. The identity of the author remains a mystery but, based on the use of certain words and sources quoted, it is reasonable to believe that the text was composed sometime between the late-eighteenth and early-nineteenth centuries. We will return to this topic later.

The existence of the text itself,[1] which is preserved in the manuscript collection of *Muséum d'Histoire Naturelle* in Paris, was first brought to the attention of Bernard Husson by his friend, Eugène Canseliet (1899–1982), the French alchemist and only direct student of Fulcanelli. This eventually resulted in the first publication of *Les Récréations hermétiques*[2] in 1964. However since the transcription by Husson turned out to be full of errors, this led another researcher, Gilles Pasquier,[3] to finally publish the first complete

---

1 MS. 362, *Les Récreations hérmetiques*, pp. 1039-1054 and the Scholies, pp. 1055-1064.
2 *Récréations hermétiques*, in *Deux traités alchimiques du XIXe siècle*, présentation et commentaire de Bernard HUSSON, Paris, L'Omnium Littéraire, 1964.
3 *L'Entrée du labyrinthe*, Éditions Dervy, 1992.

edition of the first part of the manuscript (*Les Récréations hermétiques*) together with the second part (*Scholies*) in 1992. The latter part had not been made available in the earlier edition by Husson.[4] For us, it is reasonable to consider both works as a single work, especially since they immediately follow each other in the manuscript. Even though it cannot be absolutely established as a fact that they necessarily must be the work of one and the same person, this still seems very plausible if the two texts are closely compared. As one of many examples, both texts use the exact same expression 'two flowers' (to represent Mercury and Sulphur, the white and red tinctures), for which each text provides the very same unique source, *Le Petit Paysan*, a treatise by Johann Grasshof. At the very least, it is undisputable that both texts deal with the exact same alchemical method, often using very similar expressions.

It is uncommon for a preface to an alchemical text to openly elaborate on some of the more crucial points of the work being presented, and usually the translators are

---

4   Neither did the *Scholies* appear in the later republishing by HUSSON in *Anthologie de l'Alchimie*, 1971. The many errors were not corrected. The first publishing of the *Scholies* appeared in *La Tourbe des philosophes*, issue 14–16, 1981. The present edition of the *Hermetic Recreations* is based on facsimiles of the original eighteenth-century French manuscript. A faithful transcription of the handwritten script was first made by John Koopmans, who has preserved the exact spelling of the French as set down in the manuscript. The annotations to the French transcription indicate where the more archaic conventions of the manuscript differ from modern French orthography (the modern spelling being indicated in the footnotes to the French text). The subsequent translation of the French text into English preserves the sense, style, and meaning of the original work as closely as possible. Minor adjustments have been made when it became difficult to convey the meaning of older phrases in modern English. Generally speaking, the footnotes to the English translation are provided to assist the reader in further understanding certain details of the text, or to provide additional information on some of the references cited therein.

satisfied with just presenting biographical data. In a spirit of generosity, we would like to take a slightly different approach, and openly share some of our own observations regarding the content, from an alchemical point of view.

For the reader who is already familiar with the writings of Fulcanelli, the self-proclaimed Master of Eugène Canseliet, it is fairly evident that Fulcanelli must have been very familiar with both *Les Récréations hermétiques* and the *Scholies*, even though Fulcanelli never actually mentions them in either of his two books. For example, the whole theme concerning the terms *Nostoc* and the *Flos Cœli* (Latin, 'Heavenly flower') which appears in his first work, *Le Mystère des cathédrales* (published in Paris, 1926), almost seems to echo Fulcanelli's own reading of the text. The same applies to some passages concerning the same theme in the *Hypotypose* of Pierre Dujols (1862–1926), a contemporary French Philosopher highly respected by Fulcanelli—who also does not specifically mention the text.[5] Thus, it seems likely that both these gentlemen must have considered the text as having some very secret significance. One cannot help but think that all the errors and omissions contained in Husson's edition may have been intentional, especially since he did not include the *Scholies* (perhaps the most revealing part), and since he otherwise seems to have been a very ingenious and highly educated person.

The reader who, in addition to the works of Fulcanelli, is also familiar with some of the lesser known works from the late modern period of alchemy, (18th–19th centuries), would immediately recognise the incredible similarity that exists between our translated text, the content of the German work *Wahrer alter Naturweg* (Leipzig 1782) by I.C.H.,[6]

---

5   Pierre DUJOLS, *Hypotypose*, Paris, 1914.
6   *Des Hermes Trismegists wahrer alter Naturweg. Oder: Geheimniß wie die große Universaltinctur ohne Gläser, auf Menschen und Metalle zu bereiten* (Hermes Trismegistus' True and Ancient Way of Nature. Or: Secret like the Great Universal Tincture without Glasses, for the

and also the content of the French work *Hermès Dévoilé* (1832) of Cyliani.[7]

'The anonymous author' whose work Cyliani refers to as having been printed in 'Leipzig in 1732' most likely refers to the *Naturweg* mentioned above. The confusion concerning the year of publication (1732 instead of 1782) could perhaps be explained by a misprint, where part of the number '8' was faint during the pressing, thus making it to appear as the number '3' rather than the number '8', or by a simple mistake of the typesetter, not correctly reading the date. No other text, printed in Leipzig in 1782 (nor in 1732), comes even close to such a great similarity, down to the very details, as there is between Cyliani's work and the *Naturweg*. We trust that the reader who takes the trouble to closely compare these three works will prove us right.

Cyliani exclusively describes the 'long path' (here to be understood as the usage of common gold as the source of the 'philosophical sulfur'); while the *Naturweg* describes both this long path together with the 'short path'. The latter work was also known as the 'work of the poor' (since it excludes common gold). Cyliani himself adds that he describes only the long path 'because of duty' but that he is also familiar with the short path. The latter is made with a single matter (another confirmation that the 'philosophical gold' is drawn from the matter itself and not from another matter, i.e. common gold, when following the short approach).

To this we can add that our anonymous author follows the same tradition but—contrary to Cyliani—he exclusively describes the short path. In the *Scholies*, he also makes apparent what the source of the 'gold of the philosophers'

Preparation of Metals and Men).

[7] CYLIANI, *Hermès dévoilé dédié à la postérité*, Paris, Imprimerie de Félix Locquin, 1832; Paris, Chacornac, 1915; ed. Bernard Husson, Paris, L'Omnium Littéraire, 1964.

really consists of (i.e., a by-product obtained during the preparation of the philosophical mercury through the eagles, and being called 'the ash of Hermes').

On top of this, there exists yet another very curious text, in English, from almost the exact same time as Cyliani, and which, if possible, comes even closer to our author than Cyliani or the author of the *Naturweg* does, and that is the alchemical essay by 'Zadkiel the Alchemist' a.k.a. John Palmer, entitled: 'An Essay on the Sacerdotal Science', which appears in *The Familiar Astrologer*.[8] The fact that John Palmer published his text so close to the date of Cyliani, together with the similarity in content (both use the exact same term for the menstruum, 'astral spirit'), almost makes us believe that they might have known each other, or perhaps even formed part of the same circle of alchemists or adepts. Cyliani's identity still remains an enigma, but it is known that John Palmer studied in Paris under the great chemist Louis Nicholas Vauquelin (1763–1829).[9] In fact it is in the very collection of the manuscripts that formerly belonged to the Vaquelin family (Jean Vauquelin des Yveteaux, 1651–1716) that we find our text. It includes many other interesting alchemical manuscripts, some of which have already been made available elsewhere, transcribed in French and published digitally.[10] Thus it is indeed possible to see a circle of several authors with very similar content in their texts, who could have known each other and even worked together in some capacity. The fact that the *Hermetic Recreations*, the *Naturweg*, Cyliani, and John Palmer all mention a very specific layer of three salts, formed during the preparation of the Philosophical Mercury, is another example of the remarkably strong concordance that exists

---

8   Published in 1831, see pp. 490–499, pp. 632–638 (and a beautiful and most revealing colour illustration inserted just after p. 380).

9   See Joscelyn GODWIN, *The Theosophical Enlightenment*, Albany, State University of New York Press, 1994, p.147.

10 See www.arbredor.com.

between these texts (authors). To make this explicit, let us look at the passages in question:

> A deposit of three salts occurs, namely: one that is cottony, which swims to the surface and which is the mercury; the second which is needle-like, and of the nature of Nitre, and which lies between two waters; and the third which is a fixed and mineral salt, which settles on the bottom. (*The Hermetic Recreations*).[11]

> The first salt has the aspect of wool. The second of a nitre with very fine points, and the third is a fixed, alkaline salt. (Cyliani).[12]

> [T]here will form on the top a sort of cottony mass, very brilliant and floating; this is the long desired, and much sought for Philosophical Mercury. Underneath will be found other salts, which may be brought to perfection by a continuation of the work. (Palmer).[13]

> One obtains a threefold residue of salt: (1) one which is a completely porous and fluffy salt, like fine sheep's wool; (2) a delicate nitrous salt; and (3) an alkaline salt. (*Naturweg*).[14]

---

11   See pages 66–67 of the present volume.
12   CYLIANI, *Hermès dévoilée*, Chacornac edition, p. 34: «Le premier sel a l'aspect de laine, le deuxième d'un nitre à très petites aiguilles et le troisième est un sel fixe alcalin».
13   John PALMER, *The Familiar Astrologer*, p. 635.
14   I.C.H., ed., *Des Hermes Trismegists wahrer alter Naturweg*, p. 53: „So erlanget man einer dreyfachen Ausschüsz vom [Salz], als: 1. ein ganz luckeres und wolligtes [Salz], gleich einer subtilen Baumwolle, 2. ein zart [Nitre]isches [Salz], und 3. ein alkalisches [Salz]".

All researchers of alchemical texts know very well that the greatest obstacle in trying to follow any work comes down to the actual identification of the mysterious matter(s) alluded to. This is so important that we do not even hesitate to call it the 'bottom line' of studying alchemy. Because, without this critical piece of knowledge, the texts cannot be properly understood, at least not in the way that the authors intended. Normally, this procedure of deciphering the symbols and *Decknamen* is a difficult, and often amusing process, but it seldom allows any solid conclusion. This is because alchemical texts, filled with metaphors and symbols, can be read and understood in many different ways, depending on the interpreter and his or her level of related knowledge. We do not pretend to have come up with an absolute solution in this case either, but we will candidly share some of our observations, the fruits of many years of work and study.

Let us then get straight to the point: it appears to us as if the matter alluded to by the author of the *Hermetic Recreations* is really nothing else than a type of *clay*. There are numerous examples supporting this view, and we will not wholly deprive the reader of expanding on this concept and independently making the relevant findings. Nevertheless, in order to provide some assistance, will we point out a passage from Cyliani's *Preface*:

> I must add that the matter proper to the work is that which served to form the body of primitive man: it is to be found everywhere and in varied forms; its origin is celestial and terrestrial, so too the fire of the stone.[15]

15   *Hermès Dévoilée*, Chacornac edition, p. 3: « Je dois ajouter que la matière propre à l'oeuvre est celles qui a servi à former le corps de l'homme primitif : elle se trouve partout en tout lieu, sous diverses modifications; son origine est céleste et terrestre, le feu de la pierre pareillement ».

It is well known that man was, according to the scriptures, formed from a lump or piece of clay. According to some accounts, the clay was in fact red. This is still reflected in the Hebraic name of the very first man—*Adam*—which literary means 'red' and which also recalls the denominations for 'man', 'earth', and 'clay'. René Schwaller de Lubicz (1887–1961), a French scholar and contemporary of Fulcanelli, who is sometimes alleged to have been one of the key inspirations behind Fulcanelli's books, named one of his earlier works *Adam l'homme rouge* (Adam, the red man).[16] In fact, the red coloration of clay is due to the very presence of iron. Fulcanelli, identified by some as Jean-Julien Champagne (1877–1932), but whose identity is still debated, has the following to say about clay in the *Dwellings of the Philosophers*:

> Clay owes to iron its special colouration, sometimes yellow when iron is found divided into its hydrate state, sometimes red when it is in the form of sesquioxide, a colour which is further intensified by baking (as in bricks, tiles, and pottery).[17]

16 SCHWALLER DE LUBICZ, *Adam l'homme rouge*, Officina Montalia, St. Moritz, 1927; for the alchemical contexts (and subtexts) of this text, see the recent study by Aaron CHEAK, 'The Alchemy of Desire: The Metaphysics of Eros in René Schwaller de Lubicz (A Study of *Adam l'homme rouge*)', in H. T. HAKL, ed., *Octagon*, vol. II, Gaggenau, Scientia Nova, 2016.

17 FULCANELLI, *Les Demeures philosophales et le symbolisme hermétique dans ses rapports avec l'art sacré et l'ésotérisme du Grand-Œuvre*, Paris, Jean Schmidt, 1930; éditions ALCOR, 2014, p. 55: « L'argile lui doit sa coloration spéciale, tantôt jaune quand le fer s'y trouve divisé à l'état d'hydrate, tantôt rouge s'il est sous forme de sesquioxyde, couleur qui s'exalte encore par la cuisson (briques, tuiles, poteries) ». Trans., Brigitte Donvez and Lionel Perrin, Boulder, Archive Press, 1999, p. 91, modified.

So, taking this piece of information into account, we not only have a general indication as to just any 'clay', but more specifically: a red clay, which is naturally rich in iron. Furthermore, most people who are familiar with a wide range of alchemical texts should be well aware of the close philosophical relationship and affinity which many texts establish between gold and iron. More specifically, many authors proclaim that iron itself contains the 'philosophical gold', and this to the point that it is even more abundant in iron than in gold itself (see for example Basil Valentine's *Of Natural and Supernatural Things*).[18]

It is also evident that Fulcanelli was a great supporter of this idea, which is referenced throughout his works. For example in *Le Mystère des cathédrales*, where 'Offerus' is mentioned (St. Christopher), and which Fulcanelli, on his own terms, by using the method of the phonetic cabala, interprets as 'the man who carries gold'. This is without saying that the word for iron, in French, is contained in the very name 'Offerus' (i.e. *fer*, 'ferrous').[19] There are two other curious passages in Fuclanelli's book which deserve to be highlighted, not only concerning the relevance of iron itself, but also of the relationship between clay and iron. Here are the passages (the 'second way' mentioned here refers to the 'short path', or the 'work of the poor'):

> Indubitably the artist could not pretend to acquire the original matter, that is to say, the first Adam, 'formed of red earth'; and the subject of the sages itself, qualified 'first matter of the art', is quite re-

---

18 Basilius VALENTINUS, *Von den natürlichen und übernatürlichen Dingen. Auch von der ersten Tinctur, Wurzel und Geiste der Metallen und Mineralien wie dieselben empfangen ausgekocht, geboren, verändert und vermehret werden*, in *Chymische Schrifften*, pp. 205 ff.

19 FULCANELLI, *Le Mystère des cathédrales et l'interpétation ésotérique des symboles hermétiques du Grande-Œuvre*, Paris, Jean Schmidt, 1930; éditions ALCOR, 2013, p. 41.

moved from the inherent simplicity of the 'second
Adam'. [...] The second way demands, from
beginning to end, only the help of a coarse clay
[literally: vile earth] abundantly available, of such
a low cost that in our time ten francs are sufficient
to acquire a quantity more than enough for our
needs. It is the clay [earth] and the way of the poor,
of the simple and the modest, of those whom nature fills with wonder even by her most humble
manifestations.[20]

The usage of clay in the alchemical work is generally
spoken about quite openly in the *Scholies*, the second text
which immediately follows the *Hermetic Recreations*. However, nothing is specifically revealed about the actual usage
of clay rich in iron, other than certain allusions which we
will leave for the reader to discover. We would also like to
point out that the clay referred to in the *Scholies* may not
only refer to common clay as found in the ground. Instead,
it could refer to the silty 'clay' found when a certain 'water' is fermented, putrefied, and desiccated by the ærial and
mineral Spirits. According to the *Scholies*, this 'astral Spirit'
derives from the influence of the Sun, Moon, and stars.

Before we bid the reader farewell, we would like to
highlight another curious point that binds together our

---

20   FULCANELLI, *Les Demeures philosophales*, éditions ALCOR, 2014,
pp. 109, 218: « Il est indiscutable que l'artiste ne saurait prétendre à
l'acquisition de la matière originelle, c'est-à-dire du premier Adam
« formé de terre rouge », et que le sujet des sages lui-même, qualifié
première matière de l'art, est fort éloignée de la simplicité inhérente
à celle du second Adam. [...] La seconde voie ne réclame, du commencement à la fin, que le secours d'une terre vile, abondamment
répandue, de si bas prix qu'à notre époque dix francs suffisent pour
en acquérir une quantité supérieure aux besoins. C'est la terre et la
voie des pauvres, des simples et des modestes, de ceux que la nature
émerveille jusqu'en ses plus humbles manifestations ». Trans., Donvez and Perrin, pp. 173, 331, modified.

text with those of Cyliani, Zadkiel, and even the *Naturweg*, and that is: *the whole alchemical work in question is performed, from the beginning until the end, without any usage of external fire.* According to the *Hermetic Recreations* this is openly stated (nothing other than room temperature or an ambient heat is necessary). Cyliani is a little more reserved, but still said: 'Also, consider deeply that the fire of our hearths, that of the furnaces, or a lamp is the tyrant of destruction and that nature only uses common fire in order to destroy'. In the dream allegory, when entering into the temple to combat the dragon which guards the two matters of the work (the androgynous matter as well as the astral spirit), Cyliani makes use of a lens or magnifying glass in order to heat up his weapon of attack (a spear or lance) by focusing the rays of the sun. Furthermore the Nymph (playing the role of Mother Nature and the voice of the Adept) adds: 'be very careful not to use any other fire than that of heaven'.

Bonne lecture,

CHRISTER BÖKE
& JOHN KOOPMANS

TEXTE ET TRADUCTION

TEXT AND
TRANSLATION

Récréations
Hermétiques

LES SCIENCES ÉPROUVENT COMME LES CHOSES LES vicissitudes du tems,¹ et dégénérent plutôt que d'acquérir de l'accroissement. Les hommes à Systêmes,² accueillis de toutes parts, ont semé le désordre dans le vaste champ de l'imagination, et les fleurs les plus bisarres³ en ont été le produit: ces fleurs ont pris enfin une telle faveur que les meilleurs livres, les plus beaux discours sont reputés⁴ sans valeur, s'ils n'en sont ornés.

La science dont toutes les autres dérivent, celle de la Nature, est tombée dans un tel discrédit, que l'on frappe aujourd hui⁵ de ridicule tous ceux que l'on y sait livrés.

Au moyen des lois de l'affinité, on prétend résoudre tous les problêmes;⁶ les Elémens⁷ sont ou multipliés ou anéantis; et ceux qui les admettent, sans restriction sont placés, avec ceux qui en ont traité, au rang des ignorans,⁸ ou des hommes hors de sens.

Sans repousser les affinités, bases de la nouvelle philosophie chimique, je les crois du moins inutiles au but qu'un

1. temps.
2. Systèmes.
3. bizarres.
4. réputés.
5. aujourd'hui.
6. problèmes.
7. Eléments.
8. ignorants.

HERMETIC RECREATIONS

LIKE ALL THINGS, THE SCIENCES UNDERGO THE vicissitudes of time, and tend to degenerate rather than grow.[1] Welcomed everywhere, men with structured systems have sown disorder within the vast field of the imagination, resulting in the the most bizarre flowers. These blossoms ultimately attained such favour that the best books, and the most beautiful speeches, were deemed worthless if they were not adorned by them.

The science from which all the others derive, that of Nature, has fallen into such discredit that even today one casts ridicule upon all those known to have practiced it.

By means of the laws of affinity,[2] we claim to be able to solve all problems; the Elements are multiplied or annihilated; and those who acknowledge them,[3] along with those who have written about them, are placed unreservedly into the ranks of the ignorant or senseless.

---

1. More archaically: 'acquire any accretion'.
2. Although excluded from Antoine LAVOISIER's highly influential *Traité élémentaire de chimie* (1789), affinity theories fundamentally shaped the development of the periodic table of elements. For context, see Mi Gyung KIM, *Affinity, That Elusive Dream: A Genealogy of the Chemical Revolution*, Cambridge, MA, MIT Press, 2003.
3. That is, those who acknowledge the classical, natural philosophical theory of elements as opposed to the mechanical-corpuscular theory upheld in chemistry since BOYLE's *The Sceptical Chymist* (1661).

véritable ami de la vérité se propose d'atteindre. J'entends parler ici de la connaissance des causes premiéres[9] sur lesquelles toute science doit s'assoir,[10] et qu'on affecte de mépriser, comme certain Renard de la fable, qui faisait si des raisins qu'il ne pouvait prendre: au surplus, ces lois de l'affinité que les savans[11] modernes font tant valoir, bien qu'elles ne conduisent point à la source de notre admirable fontaine de vie, sont loin d'être l'objet de nouvelles découvertes: j'en appele[12] à tous ceux d'entre eux qui ont de la bonne foi; elles étaient du moins reconnues par le fait, quand elles ne l'étaient pas encore par les mots.

Les Elémens[13] ont un CENTRUM CENTRI que tous les yeux ne peuvent appercevoir;[14] et ils ont de plus un CENTRUM COMMUNE dont les prétendus savans[15] n'osent approcher, crainte de dévoiler leur turpitude (la lumiere).[16]

Cette chaleur caustique, accompagnée de lumiére,[17] que l'on appele[18] communément feu, n'est pas l'Elément de ce nom, dont les sages ont voulu parler. On prend en cette circonstance les effets pour la cause, et on va plus loin que les rhéteurs, qui prennent au moins la partie pour le tout.

Le feu est un fluide éminemment subtil, procèdant[19] directement de la lumière et que l'on nomme, tantôt Electrique, tantôt Galvanique ou Magnètique[20] &c., suivant ses diverses modifications, ou plutôt, c'est la lumière ellemême[21] dérivée de sa source et dont elle demeure détachée. Il n'est ni froid ni chaud, et la chaleur ou le froid ne sont point des corps, quoi qu'en dise Mr. *Azaïs*,[22] mais de simples effets du mouvement ou du repos.

9. premières.
10. s'asseoir.
11. savants.
12. appelle.
13. Eléments.
14. apercevoir.
15. savants.
16. lumière.
17. lumière.
18. appelle.
19. procédant.
20. Magnétique.
21. elle-même.
22. Azais.

Without rejecting the affinities, which form the foundations of the new chemical philosophy, I believe that they are at the very least useless with regard to the goal that a true friend of truth proposes to attain. What I mean to speak of here is the knowledge of primary causes upon which any science must be established, and which one pretends to despise, like the Fox of the fable, who so regarded the grapes he was unable to reach. Moreover, although these laws of affinity that have so often been debated by modern scholars do not lead to the source of our admirable fountain of life, they are far from being the object of new discoveries. I call upon all those among them who still have some good faith; they were at least recognised by the fact,[4] if not yet by the words.

The Elements have a CENTRUM CENTRI[5] that cannot be perceived by the eye. Moreover, they have a CENTRUM COMMUNE[6] (light) which the so-called scientists do not dare to approach for fear of revealing their turpitude.

This caustic heat, accompanied by light, which is commonly called fire, is not the Element of the same name of which the sages wished to speak. In this situation one mistakes the effects for the cause, and one goes further than the rhetoricians, who at least take the part for the whole.

Fire is an eminently subtle fluid, proceeding directly from light, which is sometimes called Electric, sometimes Galvanic or Magnetic, etc., according to its various modifications. Or rather, it is light itself, derived from its source, and from which it remains detached. It is neither cold nor

---

4. That is, the fact of the existence of primary causes.
5. Latin, 'centre of the centre'.
6. Latin, 'common' or 'mutual' centre.
7. Pierre Hyacinthe AZAÏS (1766–1845), French philosopher whose most famous work was *Des Compensations dans les destinées humaines* (1809). His system claimed to explain by the law of compensations all the vicissitudes of human destinies and by the law of equilibrium all the phenomena of nature and the world.

Le mouvement seul produit la chaleur avec toutes ses conséquences bonnes ou mauvaises, ce dont chacun est en état de faire l'application; et le feu en raison de sa plus grande subtilité, est aussi le plus propre à recevoir l'impulsion et à la communiquer aux autres corps.

L'air, l'Eau et la Terre ne sont que les conséquences immédiates et successives de la formation du feu. La Lumière détachée de son foyer, accumulée par perte de mouvement et refoulée par une nouvelle et continuelle emission[23] de sa substance, s'est donnée à elle même[24] différentes formes dont nous avons fait la distinction. Dans le langage, les plus simples de ces formes ont été appelées Elémentaires.

La Lumière, principe de vie et de mouvement, peut être considérée comme l'acte unique de la création; tout le reste n'en est que la conséquence. C'est ce qu'a voulu démontrer HERMES,[25] lorsqu'il dit dans sa Table d'Emeraude: *Ce qui est dessus est semblable à ce qui est dessous, et ce qui est dessous est semblable à ce qui est dessus, pour faire au moyen de ces deux choses, le miracle d'une seule chose.*

Le Tout en toutes choses de B.V. n'est qu'une citation abrégée de cette proposition et de la vérité qu'elle renferme que tous les sages de l'antiquité ont reconnue, l'Univers signifiant l'unité retournée ou renversée en a reçu sa dénomination.

Je puis citer encore à l'appui de mon assertion, l'Evangile de S. Jean, où il est dit: *la lumière était dans les ténèbres,*[26] *et les ténèbres*[27] *ne l'ont pas comprise;* car son application morale ne fait que justifier le fait qui lui a servi de base.

---

23. émission.
24. elle-même.
25. Hermès.
26. ténèbres.
27. ténèbres.

hot, and heat and cold are not bodies, despite what Mr. *Azais*[7] says, but rather simple effects of motion or rest.

Movement alone produces heat with all its good or bad consequences, which each is capable of applying. And fire, because of its greater subtlety, is also the most capable of receiving the impulse and communicating it to the other bodies.

Air, Water, and Earth are but the immediate and successive consequences of the formation of fire. Light detached from its source, accumulated by loss of movement, and driven back by a new and continual emission of its substance, has given to itself the various forms that we have distinguished. In terms of language, the simplest of these forms were called 'Elementary'.

Light, principle of life and movement, can be regarded as the unique act of creation; everything else is but the consequence. This is what HERMES wanted to demonstrate when he said in his Emerald Tablet: *That which is above, is like to that which is below, and what is below, is like that which is above, to make by means of these two things the miracle of one single thing.*

The all in all things of B.V.[8] is but a summary of this proposition and of the truth that it contains, which all sages of antiquity have recognised; the Universe, signifying the 'returned' or 'reversed' unity, thus received its appelation.

In support of my assertion, I may also cite the Gospel of Saint John, which states: *the light shineth in darkness, and*

---

8. BASIL VALENTINE, legendary German alchemist (Benedictine Priory of Saint Peter in Erfurt, 15th century). Most researchers nowadays agree that the writings attributed to him were composed by Johann Thölde (1565–1614/1624), a German salt manufacturer who published the oldest text attributed to Valentine: *Ein kurtz summarischer Tractat, von dem grossen Stein der Uralten* (1599), possibly also influenced by other people in his circle. 'The all in all things' (*alles in allem*) corresponds to a passage from the first section of this text.

Les substances gazeuses et aëriformes[28] sont de nature cahotique[29] plutôt qu'Elémentaires, et s'invertissent[30] facilement en l'Elément dont elles se rapprochent le plus. Les Météores de toute espèce, sans excepter les aërolites[31] ou pierres d'air, prennent d'elles leur origine, cependant leur forme est toute aërienne,[32] et fait voir qu'elles sont sous la dépendance de cet Élément; mais, comme tout ce qui luit n'est pas Or, tout ce qui a la légèreté et l'apparence de l'air, n'est pas air: c'est le *Medium*[33] dont ces substances tiennent leur forme à qui cette dénomination appartient.

L'Eau, même celle des pluies et de la Rosée, n'est qu'un composé de substances gazeuses auxquelles le feu et l'action de la lumière ont donné la forme d'eau; mais c'est la forme et non la substance qu'il faut considérer ici comme Élément, or j'entends par forme ce qui en fait le lien, et qui fait aussi celui de tous les corps, même du verre.

La Terre que nous cultivons n'est pas non plus l'Elément que nous lui faisons représenter. Elle n'est au fait qu'un grand amas de débris des corps des trois Regnes[34] dans le chemin de la destruction; il est vrai de dire qu'elle contient quelques portions de la terre première et élémentaire, car indépendamment de celle que l'eau lui fournit sans cesse, elle en reprend elle même[35] la forme par sa destruction journalière. Ainsi la fin de toutes choses ressemble à son commencement et la mort devient le principe d'une nouvelle vie: c'est ce que les anciens ont reconnu et expérimenté, et qu'ils nous ont représenté sous la forme d'un serpent qui mort[36] sa queue, pour en perpétuer le souvenir.

Lors donc que vous lisez quelque traité des anciens sur l'étude de la Nature, n'entendez pas pour élément les substances crues, indigestes et mortiféres[37] que je viens de vous signaler, mais recherchez-en le *Centrum Centri* par quelques

28. aériformes.
29. chaotique.
30. s'inversent.
31. aérolithes.
32. aérienne.
33. Médium.
34. Règnes.
35. elle-même.
36. mord.
37. mortifères.

the darkness comprehended it not,[9] because its moral application only justifies the fact that served as its basis.

Gaseous and æriform substances are of a chaotic rather than Elementary nature, and easily revert to the Element to which they are closest. Meteors of all kinds, including ærolites or stones of the Air, take their origin from them, while their form is wholly ærial, which shows that they are dependent on this Element. But just as as all that glisters is not Gold, so too all that has the lightness and appearance of air is not air. Rather, it is the *Medium* from which these substances derive their form to which this name [air] belongs.

Water, even in the form of rain and Dew, is only a compound of gaseous substances to which fire and the action of light have given the form of water. But it is the form, and not the substance, which must be considered here as the Element. And yet what I understand by form is that which makes it the link, and which also constitutes the link of all bodies, even of glass.

Neither is the Earth that we cultivate the Element that we present it as. It is in fact simply a great pile of debris formed from the bodies of the three Kingdoms on the path of destruction; it is true to say that it contains a certain proportion of the primordial, elementary earth, for independently of what water constantly provides it, it resumes form by itself through its daily destruction. Thus the end of all things resembles its beginning, and death becomes the principle of a new life. This is what the ancients recognised and experienced, and what they represented to us in the form of a serpent that bites its tail in order to perpetuate its memory.

Therefore, when you read some treatise by the ancients on the study of Nature, do not understand 'element' as the

---

9. John 1:5 (KJV).

procédés[38] ingénieux et de votre propre fond;[39] car les sages le veulent ainsi, tant pour empêcher les abus, que la profanation de cette science, au moyen de laquelle la société pourrait être bouleversée et anéantie. Ne craignez donc pas de vous livrer à l'étude de notre science, et employez pour l'approfondir et en connaitre[40] les mystères, tous les efforts du raisonnement, puisqu'il n'y a que ce moyen pour sortir du Labyrinthe dans lequel vous vous êtes peut-être légérement[41] engagé. N'attendez surtout aucune preuve de nos dire,[42] car personne ne sera tenté de vous en administrer: je veux parler de cette preuve irrévocable que donne l'expérience: mais puisque d'autres l'ont acquise par les seuls moyens que je vous donne, ne désespérez pas du succès; j'ose même vous le garantir, si vous vous décidez à suivre mes conseils et à ne pas vous en écarter car je vous enseigne la droite voie et veux vous sortir des pas perdus dont la route est partout semée.

Retournez les Eléments, dit Aristote, et vous trouverez ce que vous cherchez. Cette proposition, l'une des plus importantes ayant mis les plus grands esprits en mouvement, chacun s'est mis à la recherche d'une matière première pour arriver à ce but pensant bien que les Éléments isolés ne pouvaient y conduire, tandis qu'un corps qui en était tout composé, et encore dans son état de simplicité, était le seul qu'on pouvait raisonnablement mettre en œuvre pour chercher le point de perfection. A force de chercher, quelques-uns l'ont enfin rencontré; mais ne trouvant rien dans la Nature capable de le dissoudre, malgré sa simplicité et ne pouvant en extraire les éléments par aucun autre moyen, ils s'avisèrent de remonter vers leur source commune, et et y ayant puisé, ils vinrent enfin heureusement à bout de leur dessein.

38. procédés.
39. fonds.
40. connaître.
41. légèrement.
42. dires.

raw, indigestible, and deadly[10] substances that I have just indicated to you, but rather seek it in the *Centrum Centri* by some ingenious methods of your own devising. For that is how the sages wanted it to be, as much to impede abuse as to prevent the profanation of this science, which could lead to the disruption and destruction of society. Do not be afraid, therefore, to surrender yourself to the study of our science, and to use every effort of reasoning to deepen your knowledge of the mysteries, since it is the only possible way out of the Labyrinth in which you may have become slowly entangled. Above all, do not wait for any proof of our statements, for no one will be inclined to give you any. I wish to speak of this irrevocable proof bestowed by experience, but since others have only acquired it by the means that I give you, do not despair of success. I even dare to guarantee it if you choose to follow my advice without departing from it, for I will teach you the right way because I want you to avoid the pitfalls that are sown everywhere on the path.

'Reverse the Elements', said Aristotle, 'and you will find what you seek'.[11] This proposition, being one of the most significant, and having set the greatest minds in motion, sent everyone in quest of a first matter to achieve this purpose, thinking correctly that isolated Elements could not lead to it, whereas a body that was completely composed of it, yet still in its state of simplicity, was the only one that could reasonably be used to find the point of perfection. With enormous effort, some finally discovered it; but finding nothing in Nature capable of dissolving it, despite its simplicity, and being quite unable to extract the elements by any other means, they thought it best to go back to their common source, and having drawn from it, they happily

10. Literally 'mortiferous'.
11. Literally 'return', 'reverse', or 'turn over' the elements. Although no corresponding passage in Aristotle has been identified, it should be noted that many pseudo-Aristotelian texts circulated in scholastic and esoteric contexts between antiquity and Early Modernity.

Soyez donc assuré que sans l'eau ignée composée de la pure lumière du Soleil et de la lune, il vous sera impossible de vaincre les nombreux obstacles qui se multiplieront encore à vos regards, lorsque vous tenterez le passage de ce fameux Détroit qui conduit à la mer des sages, cette eau que quelques uns[43] nomment avec raison esprit universel, et que l'Anglais Dikinson a suffisamment fait connaître, est d'une si grande vertu et pénétration, que tous les corps qui en sont touchés, retournent facilement à leur premier être.

J'ai déja[44] fait connaître que ce n'était pas l'eau de pluie ni de Rosée qui convenait à cette opération, j'ajouterai ici que ce n'est point non plus l'eau d'une espèce de champignon appelé communément *Flos Cœli* ou fleur du ciel et que l'on prend fort improprement pour le *Nostoch* des anciens, mais une eau admirable tirée par artifice des rayons du soleil et de la lune. Je dirai encore que les sels et autres aimans[45] qu'on emploie pour tirer l'humidité de l'air, ne sont bons à rien dans cette circonstance et qu'il n'y a que le seul feu de Nature dont on puisse ici se servir utilement. Ce feu renfermé au centre de tous les corps a besoin d'un certain mouvement pour acquérir cette propriété attractive et universelle qui vous est si nécessaire, et il n'y a dans le monde qu'un seul corps où il se trouve avec cette condition, mais il est si commun qu'on le rencontre partout où

---

43. quelques-uns.
44. déjà.
45. aimants.

succeeded in reaching their intended end.

Be assured, therefore, that without the igneous water composed of the pure light of the Sun and moon, it will be impossible to conquer the numerous obstacles that continue to multiply before your eyes as you attempt the crossing of this famous Strait which leads to the sea of the wise. This water, which some with good reason name the universal spirit, and which the Englishman Dickinson[12] has sufficiently made known, is of such great virtue and penetration that all the bodies which are touched by it easily return to their first state of being.[13]

I have already made it known that neither rainwater nor Dew is appropriate for this operation. I will also add here that neither is it the water of a mushroom species commonly known as *Flos Cœli* or 'flower of the sky', which is very improperly mistaken for the *Nostoch* of the ancients;[14] rather it is an admirable water drawn by artifice from the rays of the sun and the moon. I will further state that salts and other magnets, which one uses to draw moisture from the air, are good for nothing in this circumstance, and that it is only the fire of Nature that can more effectively be of service here. This fire which is contained in the center of all bodies needs a certain movement to acquire this attractive and universal property which is so necessary for you, and

---

12. Edmund DICKINSON, or DICKENSON (1624–1707) was a British alchemist mostly known from his encounter with the French adept, Theodore MUNDANUS. Dickinson is said to have witnessed several transmutations, and also published part of his correspondance with the adept; see *Epistola Edmundi Dickinson ad Theodorum Mundanum* (1686).
13. Literally 'first being'; alternatively 'first estate'.
14. *Nostoch* is term coined by PARACELSUS (1493–1541), often attributed to cyanobacteria that appeared to flourish after thunderstorms; in Mediæval times it was believed to be seeded by shooting stars (*Sternschnuppen*). See Malcom POTTS, 'Etymology of the Genus Name *Nostoc* (Cyanobacteria)', *International Journal of Systematic Bacteriology*, vol. 47, no .2, 1997, p. 584.

l'homme peut aller; c'est pourquoi j'estime qu'il ne vous sera pas difficile de le rencontrer.

M^r. Bruno de Lansac, auteur du commantaire sur l'ouvrage ayant pour titre *la lumière sortant des ténèbres*, dit savamment que le feu vit d'air et que c'est aux lieux où l'air abonde le plus qu'il faut chercher le Soufre des sages; car ils appele[46] cette eau indifféremment soufre ou Mercure, d'autant qu'elle contient l'un et l'autre et qu'elle jouit de leurs propriétés. Ce n'est cependant pas tout à fait à la lettre qu'il faut prendre ces paroles. Je recommande seulement de suivre attentivement cet auteur lorsque passant en revue les Regnes[47] de la Nature il fait une démonstration précise de l'emploi et de l'utilité de cet élément pour l'entretien de chacun d'eux. Ce chapitre bien médité sera d'un grand secours aux amateurs de la science, et je ne puis trop les engager à en faire l'objet d'une étude particulière.

J'ai dit que la lumière était la source commune, non seulement des Elémens,[48] mais encore de tout ce qui existe, et que c'est à elle, comme à son principe, que tout doit se rapporter. Le Soleil et les Etoiles fixes qui nous l'envoient avec tant de profusion en sont comme les générateurs; mais la Lune placée intermédiairement, l'attrapant de son humidité, lui communique une vertu générative au moyen de laquelle tout se régénère ici-bas.

---

46. appellent.
47. Règnes.
48. Éléments.

there is only one body in the world where it is found with this condition, yet it is so common that it is found wherever man can go. For this reason I don't think you will have any difficulty encountering it.

Mr. Bruno de Lansac, author of the commentary on the book entitled *Light Proceeding from Darkness*,[15] wisely said that fire lives by air, and that it is in the places where air is most abundant that the sulphur of the wise must be sought. This is because they indiscriminately call this water sulphur or mercury, in so far as it contains both and partakes of their properties. However, these words must not necessarily be taken literally. I simply recommend that this author be followed attentively when, reviewing the Kingdoms of Nature, he makes a precise demonstration of the function and use of this element for the maintenance of each. This chapter, well-contemplated, will be of great assistance to lovers of science, and I cannot encourage them enough to make it the object of particular study.

I have said that light was the common source, not only of the Elements, but also of everything that exists, and that it is to this, as well as to its principle, that everything must be related. The Sun and the Fixed Stars, which send this light to us in such profusion, are like its generators. But the

---

15. The *Lux obnubilata suapte natura refulgens*, Venice, 1666, was an alchemical poem published by Marc-Antoine CRASSELAME, pseudonym of the Italian marquis, Francesco Maria SANTINELLI (1627-1697); it was translated into French as *La Lumière sortant par soi-même des Ténèbres* (1687), with an extensive commentary, by Bruno de LANSAC, a man about whom little is known. Following the orders of his patron, the unfaithful translator tells us that he freely translated and edited the original text, and does not shy from declaring that he suppressed passages whenever he saw fit, 'sticking only to his spirit and intention and adding explanations'. 'I followed him scrupulously in the doctrine, but beyond that I have given it, as much as I could, a French turn, and have sought to give my translation an original air'.

Tout le monde sait aujourd hui[49] que la lumiére[50] que la lune nous envoit,[51] n'est qu'un emprunt de telle du Soleil, à laquelle vient se mêler la lumiére[52] des autres astres. La Lune est parconséquent[53] le réceptacle ou foyer commun dont tous les pphes[54] ont entendu parler: elle est la source de leur eau vive. Si donc vous voulez réduire en eau les rayons du Soleil, choisissez le moment où la lune nous les transmet avec abondance, c'est-à-dire lorsqu'elle est pleine, ou qu'elle approche de son plein: vous aurez par ce moyen l'eau ignée des rayons du Soleil et de la Lune dans sa plus grande force.

Mais il est encore certaines dispositions indispensables à remplir, sans lesquelles vous ne fairiez[55] qu'une eau claire et inutile.

Il n'est qu'un tems[56] propre à faire cette récolte des esprits astraux. C'est celui où la Nature se régénére;[57] car à cette époque l'athmosphère[58] est toute rempli[59] de l'esprit universel. Les arbres et les Plantes qui reverdissent, et les Animaux qui se livrent au pressant bésoin[60] de la génération, nous font particulièrement connaître sa bénigne influence.

Le printems[61] et l'automne sont parconséquent[62] les saisons que vous devez choisir pour ce travail; mais, le printems[63] surtout est préférable. L'été, à cause des chaleurs excessives qui dilatent et chassent cet esprit, et l'hiver à cause du froid qui le retient et l'empêche de s'exhâler,[64] sont hors de l'œuvre.

Dans le midi de la France, le travail peut être commencé au mois de mars et repris en septembre; mais à Paris et dans le reste du Royaume, ce n'est au plutôt qu'en avril

---

49. aujourd'hui.
50. lumière.
51. envoie.
52. lumière.
53. par conséquent.
54. philosophes.
55. feriez.
56. temps.
57. régénère.
58. l'atmosphère.
59. remplie.
60. besoin.
61. printemps.
62. par conséquent.
63. printemps.
64. s'exhâler.

Moon, placed intermediately, captures the light by its moisture and communicates to it a generative virtue by means of which everything here below regenerates.

Today, everyone knows that the light that the moon sends to us is borrowed from the Sun, which is in turn mixed with the light of the other stars. Consequently, the Moon is the receptacle or common source of which all the philosophers meant to speak: it is the source of their living water. Thus, if you want to reduce the rays of the Sun into water, choose the moment when the moon transmits them to us with abundance, i.e., when it is full, or when it approaches fullness. In this way you will obtain the igneous water from the rays of the Sun and the Moon at its greatest strength.

Yet there are still certain essential conditions to fulfil, without which you would only obtain a clear, useless water.

There is only one proper period to harvest the astral spirits. It is when Nature regenerates herself; for during that time, the atmosphere is completely filled with the universal spirit. The trees and Plants which grow green again, and the Animals which succumb to the compulsion to procreate, make us particularly aware of its benign influence.

Consequently, spring and autumn are the seasons that you must choose for this work. But spring above all is preferable. Summer, because of the excessive heat which dilates and drives out this spirit, and winter because of the cold which retains it and prevents it from being exhaled, lie outside the work.

In the south of France, the work may be started in March and resumed in September. But in Paris and in the rest of the Kingdom, one may only begin it in April at the

qu'on peut le commencer et la seconde sève est si faible que ce serait perdre son tems[65] que de s'en occuper en Automne.

Il faut savoir maintenant que l'influence astrale se fait préférablement sentir vers le Nord; que c'est vers le Nord que se tourne constamment l'aiguille aimantée, et que c'est aussi vers le Nord que les fluides Electrique, Galvanique et Magnètique[66] portent tous leurs efforts, c'est donc aussi vers cette région que vous tournerez votre appareil, car l'expérience a prouvé que de tout autre côté vous ne trouveriez point cet esprit.

Il faut aussi que le ciel soit pur et qu'il n'y ait point de vent, autre que la fraicheur[67] agitée de la nuit, car sans cela on n'obtiendrait qu'un esprit très faible et incapable d'action.

On peut commencer le travail aussitôt que le soleil est couché, et le continuer toute la nuit; mais, il faut le cesser lorsqu'il se lève, car sa lumière disperse l'esprit, et on ne recueille plus qu'un flegme inutile et nuisible.

Les Philosophes ont tenu jusqu'ici ces choses très secrettes;[68] ils n'en ont parlé que fort obscurement,[69] et toujours sous le voile de l'allégorie. Despagnet,[70] le Cosmopolite et quelques autres ont fait des

---

65. temps.
66. Magnétique.
67. fraîcheur.
68. secrètes.
69. obscurément.
70. D'Espagnet.

earliest, and the second sap is so weak that it would be a waste of time to attempt it in autumn.

It must now be known that the astral influence makes itself felt preferably towards the North. That is, it is towards the North that the magnetised needle constantly turns, and it is also towards the North that the Electric, Galvanic, and Magnetic fluids bring all their efforts to bear. It is therefore towards this region that you will turn your apparatus, because experience has proven that you would not find this spirit in any other direction.

The sky must also be clear and there should be no wind, other than the restless chill of the night, otherwise we would only obtain a very weak spirit incapable of action.

We can begin the work as soon as the sun has set, and continue throughout the night. But it should be ceased when it rises, for its light scatters the spirit, and we would collect nothing but a useless, harmful phlegm.

Until now, the Philosophers have kept these things very secret. They have only spoken of them in a highly obscure manner, and always under the veil of allegory. D'Espagnet,[16] the Cosmopolitan,[17] and a few others have

16. Jean D'ESPAGNET (1564–c. 1637), French lawyer and alchemist best known for his *Arcanum Hermeticæ philosophiæ* and *Enchiridion physicæ restitutæ* (1623).

17. THE COSMOPOLITAN refers to the author of *Tractatus de sulphure altero naturæ principio* (1616), i.e., Michael SENDIVOGIUS (1566–1636); there is direct influence from LIMOJON DE ST DIDIER's *Hermetic Triumph*, as we see in the following quote: 'Cosmopolite [is] more ingenious than the rest to indicate, that the Season the most proper for the philosophick Work, is that wherein all living Beings, sensitives and vegetables, appear animated with a new Fire, which carries them reciprocally to Love, and to the Multiplication of their Kinds; he says, that *Venus is the Goddess of this charming Isle*, wherein he saw naked all the Mysteries of Nature; but to denote more precisely this Season, he says, *That there were seen seeding in the Pasture, Rams and Bulls, with two young Shepherds*, expressing clearly in this witty Allegory, the three spring Months, by the three celestial Signs, answering to them, *viz. Aries, Taurus and Gemini*'.

descriptions ingénieuses de la saison du printems.[71]

Nicolas Flamel, pour désigner la région du Nord, a feint un voyage à St. Jacques de Compostelle, d'où il est revenu avec un médecin juif converti qui, après lui avoir enseigné les plus grandes particularités de l'œuvre, mourut à Orléans où il le fit enterrer à Ste. Croix.

On voit au ciel la *Voie Lactée* qui court du midi vers le Nord où elle forme deux branches dont la direction est variable en raison du mouvement de la terre, et dont la Boussole suit la variation. Cette voie lactée est appelée vulgairement le chemin de St. Jacques, parce que les Pèlerins la désignent ainsi, et qu'elle leur sert de guide pendant la Nuit pour leur grand voyage; elle est aussi le guide du pphe[72] Hermètique[73] qui la reconnaît dans le midi où elle prend sa source, et la suit vers le Nord où est son Embouchure. Le Médecin juif converti est le Mercure qu'il trouve sur sa route, et qui, comme l'on sait,[74] révèle[75] tous les secrets de l'art, quand on en est possesseur. Flamel le désigne comme médecin, parce qu'il purge les Métaux de leur lépre[76] et qu'il est vraiment une médecine. Il en est fait un juif converti, parce que la Lumière prend sa source en Orient et qu'il en fait un juste emploi. Enfin, il le fait mourir à Orléans et enterrer à Ste. Croix pour annoncer sa fixation: ce que la Croix + marquant les quatre points Cardinaux de l'atmosphère montre plus positivement. C'est donc un mensonge de l'auteur du Livre ayant pour titre *hermippus redivivus* tendant à

---

71. printemps.
72. philosophe.
73. Hermétique
74. on le sait.
75. révèle.
76. lèpre.

provided some clever descriptions of the season of spring.

Nicolas Flamel,[18] in order to indicate the area of the North, feigned a journey to Santiago de Compostela, from which he returned with a converted Jewish doctor who, having taught him the greatest peculiarities of the work, died in Orleans, where he was buried at Sainte-Croix.

In the sky we see the *Milky Way* which runs from the south towards the North, where it forms two branches, whose direction varies according to the movement of the earth, and whose variation the Compass follows. This milky way is commonly called the 'way of Saint-Jacques', because that is how the Pilgrims designate it, and because it serves as a guide for them during the Night on their long journey. It is also the guide of the Hermetic philosopher, who recognises it in the south where it draws its source, and who follows it Northward where its Mouth is. The converted Jewish doctor is the Mercury[19] that he finds on his route, and who, as is known, reveals all the secrets of the Art when it is possessed. Flamel indicates that he is a doctor because he purges metals of their leprosy, and because it really is a medicine. And he makes him out to be a converted Jew because the Light has its source in the East, and because he makes correct use of it. Lastly, he has him die in Orleans, and buries him in Sainte-Croix to herald its fixation, which the Cross, +, marking the four Cardinal points of the atmosphere, shows more distinctly. The claim by the

18. NICOLAS FLAMEL was a legendary fifteenth century French alchemist; the first alchemical text attributed to him was the *Sommaire philosophique* (Philosophical Summary) of 1561. His best-known work, *Le Livre des figures hiéroglyphiques* (The Book of Hieroglyphic Figures), was published in *Trois Traitez de la philosophie naturelle* (Three Treatises of Natural Philosophy) in 1612, together with the famous *Liber Secretus* (Secret Book) of ARTEPHIUS (c. 1150) and a text attributed to SYNESIUS. Claude GAGNON has argued that the French poet and alchemist, Béroalde de VERVILLE (1556-1626), was the true person behind all these works.

19. That is, the 'alchemical' Mercury.

accréditer son système imbécile, que la citation qu'il a fait[77] du prétendu voyage de N. Flamel et qu'il ose appuyer de la relation qui lui en fut faite par deux Adeptes se disant ses amis et affirmant sa longue existence.

B.V. fait dire à Adolphe sortant d'un souterrain à Rome, et tenant à la main le petit Coffret de plomb renfermant la figure parabolique du vieil Adam: Dans mon extrême ravissement, je regardai au midi où sont les chauds lions, et puis je me tournai au Nord où sont les Ours.

St. Didier, auteur du *Triomphe hermétique*, dans sa lettre aux disciples d'Hermes,[78] dit, que l'étude de cette Science est comme un chemin dans des Sables où il faut se conduire par l'ètoile[79] du Nord.

Cette Etoile a toujours été considérée comme le guide certain de notre pphie,[80] et c'est elle qui conduisit les bergers à la Crêche[81] où reposait le Sauveur du monde.

Il y a des ouvrages intitulés l'Étoile ou pphe[82] du Nord, mais l'abus qu'ont[83] fait de cet emblême[84] un trop grand nombre d'auteurs pseudonimes,[85] pour se donner du relief

---

77. faite.
78. d'Hermès.
79. l'étoile.
80. philosophie.
81. Crèche.
82. philosophe.
83. qu'on.
84. emblème.
85. pseudonymes.

author of *Hermippus Redivivus*[20] is thus false, for it attempts to substantiate his foolish method by citing the alleged trip of Nicolas Flamel, which he dares to support by reference to the acquaintance made between Flamel and two Adepts, thus claiming to be friends and affirming his long life.

B.V. has Adolphus say the following, as he emerges from an underground passage in Rome, holding in his hand a small leaden Casket containing the parabolic figure of the old Adam: 'In my extreme delight, I looked to the south where the ardent lions are, and then I turned to the North where the Bears are'.[21]

St. Didier, the author of the *Hermetic Triumph*, in his *Letter to the Disciples of Hermes*, said that: 'the study of this Science is like following a path in the Sand where it is necessary to be guided by the North Star'.[22]

This Star has always been regarded as the sure guide of our philosophy, and it is the one which led the shepherds to the Manger where the world's Saviour rested.

20. Johann Henrich COHAUSEN (1665–1750), *Hermippus Redivivus: Or the Sage's Triumph over Old Age and the Grave*, London, J. Nourse, 1748; without a doubt our author has at hand the French edition (1789) of the work in question which was translated from the second English edition (1748) by M. DE LA PLACE. It is noteworthy that the anecdote cited by Cohausen actually comes from a book written by the traveller Paul LUCAS, wherein a wandering dervish tells Lucas that Flamel and his wife feigned their deaths and are still very much alive, and that he last saw them in India. See *Voyage du Sieur Paul Lucas fait par ordre du Roi dans la Grece*, Nicolas Simart, Amsterdam, 1714. Anna Marie ROOS has suggested that the text is actually an iatrochemically-based satire; cf. *Medical History*, vol. 51, no. 2, 2007.
21. A reference to the dialogue between SENIOR and ADOLPHUS contained in a seventeenth century text attributed to Basilius VALENTINUS: *Occulta philosophia* (1613).
22. LIMOJON DE ST. DIDIER (c. 1630–1689), a French alchemist most famous for having written *Le Triomphe hermétique ou la Pierre philosophale victorieuse* (1690), to which the *Lettre aux vrais disciples d'Hermès* was later appended (1699).

et se faire rechercher, l'ont couvert de tant de défaveurs; qu'il a beaucoup perdu de son prix.

Sachez toutes fois[86] que l'esprit astral étant le père nourricier de la pierre, il en faut recueillir une grande quantité. Cette récolte ne peut se faire en une seule fois, c'est pourquoi on y emploira[87] tout le tems[88] que durera le travail qui est au moins de trois années; car il ne faut pas s'en tenir à ce que disent les auteurs sur les tems,[89] leurs discours n'étant que des tissus d'énigmes ou d'allégories dont je donnerai ailleurs l'explication. Revenons au principal Sujet de la Philosophie.

Tous les sages s'accordent à dire, et c'est une vérité incontestable, que l'œuvre se fait d'une seule chose à laquelle on n'ajoute rien d'étranger et dont il n'y a rien à retrancher que les immondices et superfluités. C'est ainsi que s'exprime B. Trévisan; et son dire qu'il a emprunté aux pphes[90] qui l'ont prècédé,[91] a été soutenu et répété unanimement par tous ceux qui l'ont suivi.

Bien des gens, entendant mal cette unité de la pierre, mettent dans un vaisseau qu'ils nomment un œuf pphique,[92] une seule matière de leur choix, qu'ils tiennent sur un feu de lampe ou tel autre qu'ils imaginent, et attendent ainsi vainement sa dissolution. D'autres font des amalgames, et ne sont pas mieux avisés. Ils ne font aucuns[93] progrès pour beaucoup de raisons, dont voici les principales:

86. toutefois.
87. emploiera.
88. temps.
89. temps.
90. philosophes.
91. précédé.
92. philosophique.
93. aucun.

There are works entitled the Star or philosopher of the North, but what an abuse has been made of this emblem! A great number of pseudonymous authors, in order to distinguish themselves and become sought after, have cast so much disfavour upon it that it has lost much of its value.

Know however, that because the astral spirit is the foster father of the stone,[23] it is necessary to collect a great quantity of it. This harvest cannot be made all at once. This is why we shall work on it for the entire time that the work will last, which is at least three years, for we cannot abide by what the authors say about time, since their speeches are merely the substance of enigmas or allegories, which I shall explain elsewhere. Let us return to the main Subject, which is Philosophy.

All the wise are in agreement when they say, and it is an undeniable truth, that the work is made of one single thing, to which nothing foreign is added and from which nothing is removed but the filth and superfluities. B. Trevisan expressed himself so; and his remark that he borrowed from the philosophers who preceded him has been unanimously supported and repeated by all those who followed him.

Many people, misunderstanding this unity of the stone, place a single matter of their choice into a vessel which they call a philosophical egg; they keep it over a lampfire or such other as they fancy, and thus wait in vain for its dissolution. Others make amalgams, and are no better advised. They make no progress for numerous reasons, principal among which are:

---

23. Literally the 'nourishing' or 'feeding' father (*père nourricier*), with a possible play on words between *père* (father) and *pierre* (stone).

1º Ils travaillent sur une matière morte; et quand ce sera sur le véritable sujet de la pphie,[94] le vase et le feu ne lui sont pas proportionnés.

2º Ils ignorent que depuis le commencement jusqu'à la fin du travail, notre matière est double, je veux dire qu'elle a un agent et un patient sans lesquels il n'y aurait aucune action dans le vaisseau: que l'agent fait office de mâle, et le patient celui de femelle, et que tous les deux ensemble, bien que séparés par leur Nature, ne constituent qu'un seul corps qui est nommé à cet effet *Rebis* ou deux choses en une.

3º Enfin, leur travail est tout à fait en sens inverse de celui de la Nature; car ils ne savent ni dissoudre, ni putréfier, ni distiller, ni sublimer, ni aucune de nos opérations. Cependant ils ne laissent pas d'entreprendre, se disant à eux mêmes[95] : cet œuvre est celui de la Nature à qui nous n'avons besoin que de prêter la main, c'est à elle de l'achever. Marchant ainsi en aveugles, et avec tant de confiance, ils ne peuvent manquer de se heurter à chaque pas qu'ils font dans un si obscur dédale.

Nous lisons dans l'Evangile qu'il ne vient pas de Lys sur des Ronces, ni de figues à la place de raisins; que telle est la semence, et tel sera le fruit; mais qu'un mauvais arbre ne peut produire de bons fruits, et que, pour cela, il doit être coupé et jetté[96] au feu; mais ces raisons ne les touchent point, et ils n'en sont pas moins persuadés de réussir. Cependant voyant la mauvaise fin de leur travail, ils devraient s'amender et reconnaître leur faute; mais, bien loin de là, ils l'attribuent

---

94. philosophie.
95. eux-mêmes.
96. jeté.

1. They work on dead matter. And even when it is the true subject of philosophy, the vessel and the fire are not proportionate to it.
2. They are unaware that from the beginning to the end of the work, our matter is double. I mean that it has an agent and a patient, without which there would be no action in the vessel; that the agent serves as the male and the patient as the female; and that both together, although separated by their nature, constitute but one single body, which for this purpose is called *Rebis*, or two things in one.
3. Finally, their work runs completely counter to that of Nature, because they do not know how to dissolve, putrefy, distil, sublimate, or perform any of our operations. However they do not refrain from undertaking it, saying to themselves: this work is that of Nature to whom we only need to lend a hand, it is for her to complete it. Proceeding thus in blindness, yet with such confidence, they cannot avoid stumbling with each step that they take in so dark a maze.

We read in the Gospel that the Lily does not come from Thorns, nor figs from grapes; as is the seed, so shall be the fruit. But a bad tree cannot produce good fruits, and because of this, it must be cut down and cast into the fire.[24] But these reasons do not affect them, and they are no less persuaded to succeed. However, seeing the bad result of their work, they should make amends and acknowledge their mistake. But, far removed from this, they attribute it to some unforeseen accident and return with renewed cour-

---

24. Luke 3:9; 6:43–44; Matthew 7:18–19.

à quelque accident qu'ils n'ont pu prévoir, et se remettent avec plus de courage encore à leur sot ouvrage. Mais, laissons ces ignorans[97] s'enfler à loisir de vaines fumées et ne nous occupons plus que du choix d'une matière due et de sa préparation.

Il s'agit moins de passer en revue les substances des trois Règnes, que d'examiner leur composition, pour savoir de quoi elles ont été formées. A la première vue, cette difficulté parait[98] insurmontable. Elle est grande, à la vérité, mais pas autant qu'on pourrait se l'imaginer; car,

1° Nous n'avons besoin pour ce travail, ni d'Alambic, ni de Cornues, encore moins de Sels, d'Esprits ardents, acides ou Corrosifs &c.
2° Nous savons au surplus que toutes les choses de ce monde ont une même origine, et qu'elles ne diffèrent entre elles que par le mêlange[99] des Elémens,[100] mais tels que je les ai dépaints[101] plus haut.
3° Il ne nous reste plus en troisième lieu qu'à rechercher exactement le point de leur formation.

Considérez que le Ciel et la Terre ont premièrement existé; que le Ciel servant d'agent ou de mâle, et la terre de patient ou de femelle ont donné naissance à toutes choses. Cependant ils n'étaient pas distincts l'un de l'autre, et ils ne formaient d'abord qu'une masse ténébreuse et abominable; mais la lumière en ayant été séparée, et les cieux en ayant été établis, la masse s'ébranla et donna signe de vie. Les Elémens[102] furent formés, l'Univers et tout ce qu'il renferme parut ensuite; et cet ordre si admirable de choses subsiste dépuis[103] cette époque, et demeurera ainsi jusqu'à ce qu'il plaise au Souverain Médiateur de le changer.

97. ignorants.
98. paraît.
99. mélange.
100. Éléments.
101. dépeints.
102. Éléments.
103. depuis.

age to their foolish work. But let us leave the ignorant to inflate themselves at their leisure with empty smoke, and no longer concern ourselves with anything but the choice of one proper matter and its preparation.

It is less a question of reviewing the substances of the Three Kingdoms, than to examine their composition in order to know what they were formed from. At first sight, this difficulty appears insurmountable. It is, in truth, great, but not as much as we might expect, because:

1. For this work, we need neither an Alembic, nor Retorts, and even less Salts, ardent Spirits, acids or Corrosives, etc.
2. Furthermore, we know that all things of this world have the same origin, and that they differ only by the mixture of Elements, as I have related above.
3. In the third place, the only thing that remains for us to do is to seek the exact point of their formation.

Consider that the Heavens and the Earth existed first—that the Heavens, serving as the agent or male, and the Earth, as patient or female, gave birth to all things. Originally, however, they were not distinct from each other, and at first they only formed a tenebrous, abominable mass. But when the light was separated from this mass, and heaven was established, the mass shook and gave signs of life. The Elements were formed, and then the Universe and all that it contains appeared. And such an admirable order of things has remained ever since, and will continue to remain until it pleases the Sovereign Intermediary to change it.

La vie telle qu'on voudra la considérer, n'est qu'un combat de deux substances, ou un échange continuel de lumière et de ténèbres,[104] l'une de ces substances prend alternativement la place de l'autre, et fait tantôt fonction de mâle et tantôt de femelle; de manière que quand il plaît au divin auteur, tout se change en une pure lumière ou tout retourne dans les ténébres[105] cineriennes,[106] ce qui fait voir que la lumière et les ténébres[107] ne sont qu'une même chose, changée de forme et de valeur par le développement ou le resserrement de la substance, que de là provient un attrait mutuel d'où ressort, avec le mouvement, l'inversion élémentaire de la substance.

Qui habet aures audiendi, audiat.

Considérez maintenant que de la même manière et de la même matière dont le monde a été créé, l'œuvre des sages est mis aujourd,[108] et que c'est pour cette raison qu'il a reçu le nom de petit monde ou Microcosme. Ainsi, je vous ai dit en peu de paroles tout ce que vous avez à faire pour cette grande entreprise.

Prenez donc la terre première qui n'est qu'une pure lumière environnée de ténèbres,[109] et réduisez la[110] en ses principes avec la pierre arrachée sans mains du sommet de la montagne, afin de reconnaitre[111] en elle trois substances distinctes qui sont le sel, le soufre et le Mercure, lesquelles étant adroitement conjointes avec les deux dont la matière est formée, à savoir le Ciel et la Terre, forment une Quintessence admirable dont les vertus sont infinies et incompréhensibles.

Cette pierre merveilleuse apparut en songe à Nabuchodonosor Roi de Babylone, et vint briser et réduire en pou-

104. ténèbres.
105. ténèbres.
106. cimmériennes.
107. ténèbres.
108. au jour.
109. ténèbres.
110. réduisez-la.
111. reconnaître.

Life, such as we wish to consider it, is but a struggle between two substances, or a continual exchange of light and darkness. One of these substances alternatively takes the place of the other, sometimes taking the male function and sometimes the female. And in a manner pleasing to the divine author, everything either changes into a pure light, or returns to the Cimmerian darkness, which shows that light and darkness are but one and the same thing, changing in form and value by the expansion or contraction of the substance. And from this comes a mutual attraction whence, with movement, the elementary inversion of the substance arises.

Qui habet aures audiendi, audiat.[25]

Consider now that in the same manner, and from the same matter of which the world was created, the work of the wise is revealed, and that is why it has been called the 'small world' or Microcosm. Thus, in a few words I have told you everything that you need to do in order to realise this grand undertaking.

Take, therefore, the first earth which is but a pure light surrounded by darkness, and reduce it into its principles with the stone wrenched without hands from the mountain summit, in order to recognise three distinct substances in it: salt, sulphur, and Mercury. These three are skillfully conjoined with the two from which the matter is formed, namely Heaven and Earth, to form a remarkable Quintessence whose virtues are infinite and incomprehensible.

This marvelous stone appeared in a dream by Nebuchadnezzar, King of Babylon; it came to break down and reduce to powder a large statue which he saw standing be-

---

25. Mark 4:9 (Vulgate): 'He that hath ears to hear, let him hear' (KJV).

dre une grande statue qu'il voyait debout devant lui, et dont la tête était d'or le plus pur, la poitrine, les épaules et les bras d'argent le ventre et les cuisses d'airain, les jambes de fer, et l'argile y était amalgamée avec de la semence humaine, mais qui ne leur était point adhérente, non plus que le fer ne peut être mêlé avec l'argile.

Nabuchodonosor justement effrayé de cette vision, manda tous les mages de son Royaume, et exigea deux,[112] sous peine de mort, qu'ils devinassent son songe et en donnassent une juste interprétation; aucun d'eux n'en put venir à bout. Il n'y eut dans tout le Royaume qu'un jeune homme nommé Daniel, et rempli de l'esprit de Dieu, qui pût satisfaire à sa demande. Daniel chap. 2. v. 18.

Ce songe peut être appliqué tout entier à l'Œuvre des sages, et lui servir de figure Parabolique. On verra, par exemple, dans les Mages de Babylone, la Tourbe des faux savans[113] s'efforçant en vain d'entendre la science, voulant néanmoins persuader qu'ils la possèdent, et conduisant dans des sentiers perdus ceux qui se livrent à eux de trop bonne foi: dans Daniel, un fils de la sagesse, à qui tous les secrets de la Nature sont connus, et qui peut en donner une seine[114] et véritable explication.

La statue sera notre Arbre Métallique depuis son sommet jusqu'à sa racine dans laquelle sont encore confondus, Saturne, Jupiter et Mercure comme métaux de première origine. Le fer et l'argile mêlés avec de la semence humaine représenteront l'Œuvre de Nature figuré de main d'homme; et la pierre coupée sans mains du haut de la montagne, et venant briser les pieds de la statue et la réduire en poudre impalpable sera prise ou pour la foudre que lance Jupiter, ou pour la faulx[115] de Saturne que vous devez échanger

---

112. d'eux.
113. savants.
114. saine.
115. faux.

fore him, whose head was of the purest gold, whose chest, shoulders and arms were of silver, whose belly and thighs were of copper, whose legs were of iron and of clay combined with human seed, which adhered no more than iron mixed with clay.

Nebuchadnezzar, rightfully troubled by this vision, summoned all the Magi of his Kingdom, and under penalty of death, demanded them to divine his dream and provide a correct interpretation of it. But none of them were able to do so. In the entire Kingdom there was only one young man who could satisfy his request; his name was Daniel and he was filled with the spirit of God (Daniel chapter 2, verse 18).

This dream can be wholly applied to the Work of the wise, and serves as a Parabolic Figure. We shall see in the Magi of Babylon, for example, an assembly of false scholars trying in vain to understand the science, yet nevertheless wishing to persuade others that they possess it, while leading astray those who place too much good faith in them; in Daniel [we see] a son of wisdom to whom all the secrets of Nature are known, and who can give a healthy and true explanation.

The statue represents our Metallic Tree from its summit to its root, in which Saturn, Jupiter, and Mercury, as metals of first origin, are still confounded. Iron and clay mixed with human seed represent the Work of Nature rendered by the hand of man. And the stone cut without hands from the mountain top, which comes to break the legs of the statue to reduce it to an intangible powder, is to be taken for either the lightning which Jupiter hurls, or for the scythe of Saturn that you must skillfully exchange for the trident of Neptune,

adroitement contre le Trident de Neptune, moyennant une certaine clef qui je vous donnerai, jusqu'à ce que Pluton s'en montrant jaloux, et soufflant du fond de ses caverns, montre à son tour sa puissance, en desséchant les eaux, et réduisant l'arbre en cendre ou poussière que vous sèmerez et dont il viendra beaucoup de pierres précieuses.

Les Anciens, jaloux de leur secret, ont parlé de la matière sous ses divers aspects, afin de tromper la crédulité des gens avares et des ambitieux qui ne rêvent que de puissance et dévastation. Ils ont confondu avec le sujet de la p.phie[116] leur première matière qui ne s'obtient qu'après beaucoup de tems[117] et de longs travaux. N'étant nullement participant de leur envie, j'ai voulu vous faire toucher du doigt ce sujet tant recherché et l'ai mis exprès tout nud[118] devant vos yeux, pour vous dispenser de le chercher plus longtemps. J'espère que vous me saurez gré de ma franchise, et que vous en tirerez le parti le plus avantageux, en vous prevenant[119] toutefois d'ajouter à mes paroles un petit grain de sel, pour vous les rendre plus sensibles.

Ferrare peint ce sujet comme une pierre qui n'est pas pierre, qui est dure et molle, et qui n'est d'aucun prix; mais si vous voulez m'en croire vous vous attacherez davantage à ce qu'en a dit le comte de Trévise, car il s'est montré moins envieux que personne, ayant peint ce sujet très au long dans son *Arca Aperta*, et ayant fait une description très étendue des matières qui ne sont pas propres à notre Œuvre, dans un autre ouvrage. Je vous donnerai ensuite le conseil de l'illustre commentateur de la *Lumière sortant des ténébres*,[120]

---

116. philosophie.
117. temps.
118. nu.
119. prévenant.
120. ténèbres.

by means of a certain key that I will give you, until Pluto, revealing his jealousy of it, and blowing from the depths of his caverns, in turn demonstrates his power by desiccating the waters and reducing the tree to an ash or dust that you will sow, and from which many precious stones will arise.

The Ancients, possessive of their secret, have spoken about the matter under diverse aspects in order to misdirect the credulous nature of the avaricious and ambitious people who only dream of power and devastation. They have obscured their first matter, which is only obtained after much time and long work, with the subject of philosophy. Not sharing their possessiveness at all, however, I wanted to let you grasp this ever-sought-after subject, and so I have expressly laid it bare before your eyes in order that you may dispense with your seeking. I hope that you will be grateful for my frankness, and that you will derive the most advantage from it. But be sure to take my words with a small grain of salt in order to render them more sensible.

Ferrare[26] paints this subject as a stone which is not a stone, which is hard and soft, and which is of no value. But if you want to believe me, you should become more familiar with what the Count of Trevise[27] said, because he has shown himself to be less possessive than anybody, having painted this subject at great length in his *Arca Aperta*, and having made an extensive description of the matters which are not specific to our Work, in another work. I will give you then the advice of the illustrious commentator of the *Light departing from darkness*, Mr. Bruno de Lansac: 'Choose', he

---

26. Likely refers to PETRUS BONUS of Ferrara (fl. ca. 1323-1330), an Italian alchemist most known for his *Margarita Preciosa*, first printed in 1572 but circulated in manuscripts long before.
27. BERNARD TREVISAN, Italian alchemist, c. 1406-1490. He is mostly known for his *Trevisanus de Chymico miraculo, quod lapidem philosophiae appellant* (1583) featuring his well-known fountain allegory, much quoted by later alchemists. Notice that the author of the present work here confuses Bernard Trevisan with Johann GRASSHOFF's *Arca Aperta*. See note 44.

Mr. Bruno de Lansac : choisissez, dit-il une matière qui ait le brillant métallique, et j'y ajouterai qu'elle ne soit point métal, ni minéral, autrement elle ne servirait de rien. Vous saurez au surplus que ce brillant n'est que le cachet de la matière et ce qui la décèle aux yeux du sage, et vous prendrez garde de prendre le fruit au lieu de la racine; car non seulement il est immur,[121] mais dans une hypothèse opposée, il ne vous donnerait encore qu'un sauvageon dont vous ne tireriez aucun parti.

La dissolution est la première chose qu'il vous faut entreprendre, car il faut délier le corps pour mettre les ennemis aux prises. Or le feu et l'eau vous seront ici grandement nécessaires, d'autant que ces élémens[122] sont déjà ennemis de leur Nature et ne demandent qu'à essayer leurs forces.

L'esprit, dont je vous ai parlé plus haut, est un feu vaincu par l'eau dont vous vous servirez à cet effet. Vous en emplirez le Vase de Nature et vous le distillerez à feu très lent pour le déflegmer. Vous trouverez au fond quelque chose de fixe que vous vous garderez d'en retirer. Vous verserez dessus de nouvel esprit dans la même proportion, et vous continuerez ainsi les distillations, jusqu'à ce que le vase n'en puisse plus contenir, et que tout demeure fixe au fond. En continuant le feu au même degré, vous apercevrez bientôt dans votre vaisseau quelqu'agitation[123] causée par un petit vent de Sud-ouest, laquelle sera suivie d'une pluie fort agréable à la vue. Le vent et la pluie allant toujours croissant, vous ne verrez plus dans le vaisseau que comme une mer qui sera de plus en plus agitée jusqu'à ce qu'enfin les élémens[124] pacifiés, tout rentre dans l'ordre de la Nature. Mais le jour a fait place à la Nuit, l'obscurité s'agrandit et le vaisseau est d'un noir parfait. Cette Nuit est la cinquantième, et elle a paru

---

121. non mûr.
122. éléments.
123. quelque agitation.
124. éléments.

says, 'a matter which has the metallic lustre', and I shall add that it is neither metal, nor mineral, otherwise it would be useless. You will know, moreover, that this lustre is but the seal of the matter, which reveals it to the eyes of the wise, and you must be careful to take the fruit instead of the root, because not only is it unripe, it is oriented in the opposite direction,[28] and would still only yield a wild offshoot from which you would gain no use.

Dissolution is the first thing that you must undertake, for the body must be unbound in order to allow the enemies to engage. Now, fire and water will be much needed here, especially since these elements are already inimical by Nature, and are eager to test their strength.

The spirit, of which I have spoken above, is a fire that is conquered by water, which you will use for this purpose. Fill the Vase of Nature with it and distil it on a very slow fire to dephlegmate it.[29] You will find at the bottom something fixed that you must be careful not to withdraw. Pour new spirit on top of it in the same proportion, and continue the distillations until the vase cannot hold any more, and everything remains fixed at the bottom. While maintaining the same degree of fire, you will soon see some strong agitation in your vessel, which is caused by a light south-westerly wind, which will be followed by a rain that is very pleasing to the eye. The wind and rain will continue to increase until you are no longer able to see anything in the vessel but a sea, which will become more and more agitated, until at last the elements become pacified, and everything returns to the order of Nature. But day gives way to Night, the darkness increases, and the vessel becomes a perfect black. This is the fiftieth Night, and to the sailors it seemed to be thrice

---

28. Literally 'in an opposed hypothesis' (*hypothèse*).
29. *Déflegmer* (from *flegme*, 'phlegm, composure'); perhaps therefore 'decompose'.

triple aux matelots à cause de la fatigue qu'ils ont essuyée. Le jour commence à poindre, l'horison[125] est clair et sans nuage; la journée sera magnifique.

Cette manière de s'exprimer est commune à presque tous les auteurs anciens, et il n'est pas rare de trouver des lecteurs qui prennent ces discours à la lettre. Le vent et la pluie sont pour eux des réalités, et leur crédulité embrasse les plus petits détails de l'allégorie. Celle-ci que je vais remettre dans le sens droit leur facilitera l'intelligence des autres.

Le Vase de Nature est ici la terre préparée qu'il faut abreuver de son esprit. Elle est dite un Vaisseau, et elle l'est en effet, puisqu'elle contient. L'Esprit qu'on lui ajoute n'est point une chose étrangère puisque tout est sorti de lui, et que notre terre en est formée; c'est pourquoi il est dit de faire rentrer l'enfant dans le ventre de la mère: ce qui ne peut se faire qu'en lui déchirant les entrailles. Il faut aussi que notre terre soit divisée dans ses plus petites parties pour mettre au jour ses grandes richesses, et cela arrivera ainsi, si vous l'abreuvez souvent de son esprit et que vous la laissiez autant de fois dessécher. Dans cette opération, le flegme s'évapore, mais l'esprit demeure et s'incorpore avec la terre qu'il salifie jusqu'à ce que la saturation soit complette[126] ; alors l'esprit qu'on ajoute ne pouvant plus être contenu réagit sur celui que la terre a fixé et l'oblige à se dissoudre, ainsi que farait[127] le sel; pourquoi cette dissolution est comparée à une mer; et parce que l'esprit qu'on ajoute est joint à une humidité altérante et corrompante, il résulte de son mêlange[128] un mouvement de fermentation qui est suivi de putréfaction, et par conséquent de régénération, parce que la fermentation change les corps de Nature, et dans la putréfaction, ils ne font qu'échanger leurs vêtemens[129] contre de

125. l'horizon.
126. complète.
127. ferait.
128. mélange.
129. vêtements.

that amount due of the fatigue that they endured. The day begins to dawn, the horizon is clear and cloudless; the day will be magnificent.

This manner of expression is common to almost all of the ancient authors, and it is not uncommon to find readers who take these discourses literally. The wind and rain are realities for them, and their credulity embraces the smallest details of the allegory. This allegory, which I will set straight, will facilitate the understanding of others.

Here the vase of Nature is the prepared earth, which must be watered with its spirit. It is said to be a vessel, and indeed it is, for it 'contains'. The spirit that is added to it is by no means something foreign, since everything originates from it, and because our earth is formed from it. This is why it is said to 'make the child return to the womb of the mother' —which can only be done by tearing out her entrails. It is also necessary that our earth be divided into its smallest parts in order to bring to light its great riches, and this will happen if you water it frequently with its spirit, and let it dry out just as often. In this operation, the phlegm evaporates, but the spirit remains and is incorporated with the earth which it salifies, until the saturation is complete. Then the spirit which is added, which can no longer be contained, reacts upon that which was fixed by the earth, forcing it to dissolve, as salt does. This is why this dissolution is compared to the sea. And because the spirit which we add is joined to a corrosive[30] and corrupting humidity, the mixture results in a process of fermentation, followed by putrefaction, and consequently regeneration. This is because fermentation changes Nature's bodies, and in the putrefaction, they only exchange their vestments for new,

---

30. French *altérante* (altering) generally has negative connotations ('distorting, corroding, corrupting') but may also mean 'thirsty'.

nouveaux et d'autant plus riches et brillans,[130] que l'Esprit moteur est d'une origine plus relevée.

Ce que la Matière peut contenir d'humidité, sans la déverser en dihors,[131] voila[132] la mesure à observer pour les imbibitions, et ce que nous appelons le poids de Nature.

La matière servant de vase, sert également de fourneau, puisque l'esprit que vous y introduisez est un feu naturel qui la cuit et la digère pour me servir, jusqu'au bout, des expressions p.phiques.[133]

Il ne faut pas moins de cinquante ablutions; car chaque ablution jusqu'à la parfaite dessiccation, est comptée pour un jour naturel ou p.phique[134] ; de manière que nos jours peuvent durer une semaine, suivant la saison, la qualité et la quantité de matière soumise au travail. Le grand secret des Sages pour abréger le temps, est de diviser la matière, pour que les jours aient moins de longueur.

Quoique nous ne nous servions point de feu vulgaire pour nos opérations, il est néanmoins certain que nous avons besoin d'une température assez élevée pour que l'évaporation puisse se faire et que la matière ne languisse pas, et ne se perde. Il est par conséquent utile et indispensable, pendant l'hiver, et dans le lieu de travail, de faire un peu de feu, mais non assez pour que la matière en soit échauffée, ce qui serait pis que de n'en point avoir; parce que l'esprit serait chassé et ne pourrait être remplacé. Il ne faut pas que la température passe quinze degrés de Réaumur.[135]

Lorsqu'on a ainsi opéré et que la matière se dissout, elle noircit à mesure. On ne lui ajoute dans ces divers tems que l'esprit nécessaire pour entretenir son feu fermentatif; et quand la matière commence à fermenter, il faut l'aban-

---

130. brillants.
131. dehors.
132. voilà.
133. philosophiques.
134. philosophique.
135. 19° Celcius.

richer, and more lustrous ones, because the driving Spirit is of a higher origin.

What the Matter can absorb of humidity, without letting it leak out, is the measure to observe for the imbibitions, and this is what we call the 'weight of Nature'.

The Matter used as a vase serves equally as a furnace, since the spirit that we introduce to it is a natural fire, which cooks and digests it, and serves us until the end of the philosophical articulations.

No less than fifty ablutions are required, for each ablution, until the perfect desiccation, counts as one natural or philosophical day; in this way, each of our 'days' can last a week, depending on the season, as well as on the quality and quantity of matter subjected to the work. The great secret of the Wise for shortening the time is to divide the matter so that the days are less long.

Though we do not use vulgar fire for our operations, it is nevertheless certain that we need a high enough temperature so that evaporation can take place, and so that the matter does not languish or get lost. In the workplace, during winter, it is consequently useful and indispensible to make a small fire, but not so much that the matter is overheated by it, which would be worse than having no fire at all, because the spirit would be driven out and could not be replaced. The temperature must not exceed fifteen degrees of Reaumur.[31]

When we operate in this manner, and when the matter dissolves, it darkens by degree. At these different stages, we only add the spirit necessary to maintain its fermentative fire. And when the matter begins to ferment, we must

---

31. The equivalent of about 19° Celsius (66° Fahrenheit).

donner à son propre feu, jusqu'à la blancheur parfaite où elle arrive d'elle-même.

La matière n'est pas liquide comme un Brouët,[136] mais épaisse et noire comme de la poix ou du cirage de bottes; elle se boursouffle,[137] s'élève dans le Gobelet, donne des Bulles quel'on[138] compare à des yeux de poissons et qu'il ne faut pas crever, car elles contiennent l'esprit animateur.

Après la fermentation, la matière s'affaisse; elle est alors luisante comme de la poix, et du plus beau noir; c'est le signe de la putréfaction que l'on nomme tête de corbeau. Elle se dessèche ensuite peu à peu et passe à la couleur gris de cendres. Bientôt un cercle Capillaire de la plus éclatante blancheur paraît autour du vaisseau. Ce Cercle s'élargit de plus en plus jusqu'à ce que le tout soit d'une blancheur parfaite.

Avant que cette blancheur arrive, il paraît[139] quelques couleurs sur la matière, parmi lesquelles domine la verte, mais elles ne sont pas très prononcées, et ne sont que passagères et de peu de durée. On les compare néanmoins à l'Iris ou arc-en ciel.[140] Ce n'est que dans les opérations subséquentes qu'elles ont un caractère très prononcé.

Vous avez passé en revue, sans vous en apercevoir, nos différentes espèces de feux, le premier, jusqu'à la fermentation, est appelé Bain marie, ou de mer, parce qu'il n'opère, en quelque façon, qu'une dissolution saline. Le second est appelé chaleur de fumier, et vous en savez maintenant la raison. Le troisiéme[141] est appelé feu de cendres; et le quatriéme[142]

---

136. Brouet.
137. boursoufle.
138. que l'on.
139. paraît.
140. arc-en-ciel.
141. troisième.
142. quatrième.

abandon it to its own fire, so that it comes to the perfect whiteness by itself.

The matter is not liquid like a Broth,[32] but thick and black like pitch or boot polish. It swells, rises in the Beaker, and produces Bubbles which are comparable to fish eyes, and which should not be burst, for they contain the animating spirit.

After the fermentation, the matter subsides. It then shines like pitch, and is the most beautiful black. This is the sign of putrefaction known as the raven's head. It then dries out little by little, and turns an ash-grey colour. Soon, a Capillary Circle[33] of the most dazzling whiteness appears around the vessel. This Circle widens more and more until the whole [surface] is perfectly white.

Before this whiteness appears, some colours appear on the matter, among which green predominates. However, they are not very pronounced, and are also fleeting and short-lived. Nevertheless they can be compared to the Iris or rainbow. It is only in subsequent operations that they have a more pronounced character.

You have passed in review, without noticing it, our different kinds of fire. The first, up until fermentation, is called the Bain-marie, the 'bath of the sea', because in some way it only operates as a saline dissolution. The second is called the heat of manure, and now you know the reason. The third is called the fire of ashes. And the fourth, finally, is

32. The French word *Brouet* in this context probably refers to the *Brouet noir*, the 'black soup' or 'broth' (μέλας ζωμός) with which the Spartans nourished themselves.

33. Literally 'hairlike circle' or 'ring' (*cercle Capillaire*), referring to FLAMEL's Capillary Whiteness, so-called because at the end of the regime of Jupiter little filaments appear like white hair. Cf. FULCANELLI (*Demeures Philosophales*): 'Artephius, Nicolas Flamel, Philalethes, and many other masters teach that, at this stage of the coction, the Rebis takes on the appearance of fine silky threads, of *hair* spread on the surface and progressing from the periphery to the centre. Hence the name capillary whiteness used to refer to this coloration'.

enfin feu de Réverbère. Nous avons encore d'autres espèces de feux, mais qui connait[143] les premiers, connait[144] indubitablement tous les autres. D'ailleurs nous les signalerons au passage.

Vous remarquerez ici que ce travail ressemble à celui des jardiniers qui arrosent leurs jardins. Qu'arrive-t-il en cette circonstance ? la terre végétale qui, comme je vous l'ai faite observé dès le commencement, n'est formée que de débris des corps, s'altère et se décompose par sécheresse et humidité successives, et fournit un sel et un esprit dont la plante se nourrit par le moyen de l'eau qu'elle absorbe et qui est le conducteur.

Je reviens à la matière blanchie et qui est encore bien éloignée du but où vous devez la conduire. Néanmoins la principale serrure est ouverte, il n'y a plus qu'à pénétrer dans le sanctuaire, mais toujours avec précaution pour ne point faillir, et être obligé de s'arrêter en si beau chemin.

Cette poudre blanche ou matière régénérée est le Mercure encore enfant, et à qui il faut donner des ailes d'aigle à la tête et aux talons, c'est-à-dire depuis les pieds jusqu'à la tête, pour qu'il puisse voler, et s'élever à la plus haute région qui est le Ciel. Il faut le sublimer autant de fois que dans sa dissolution dans l'esprit astral, il laissera une terre en arrière qui se précipitera et qu'il vous faudra recueillir avec beaucoup de soin. Philalette[145] appele[146] ces sublimations des aigles; d'autant que le mercure acquiert chaque fois une grande subtilité, et il compare la terre que le Mercure jette en arrière, à la queue que laisse le mercure vulgaire derrière lui, tant qu'il n'est pas assez purifié. Lavez, dit-il votre mercure et le purifiez par sel et vinaigre, jusqu'à ce qu'il ne laisse plus de queue derrière lui, en coulant sur une surface plane.

143. connaît.
144. connaît.
145. Philalèthe.
146. appelle.

called the fire of the Reverberator. We still have other kinds of fires, but whoever knows the first, undoubtedly knows all the others. In any case, we will point them out as we go.

You will notice here that this work resembles that of gardeners watering their gardens. What happens in this circumstance? The vegetal earth, or topsoil, which as I have made you observe from the beginning, is only formed from the debris of bodies, has deteriorated and decomposed by successive periods of dryness and humidity; it provides a salt and a spirit which the plant nourishes itself on by means of the water that it absorbs, which is the conductor.

I now return to the whitened matter, which is still quite far from the goal to which you must carry it. Nevertheless, the principle lock having been opened, all that remains is to penetrate the sanctuary, but always with sufficient precaution to avoid failure, so you are not forced to stop along such a beautiful path.

This white powder, or regenerated matter, is Mercury, who is still a child, and whose head and heels must be given eagle wings, i.e. from the feet to the head, so that he can fly and reach the highest region, which is the Sky (Heaven). It should be sublimated as many times as its dissolution in the astral spirit; it will leave an earth behind which will precipitate and which you will have to collect with the utmost care. Philalethes calls these sublimations the eagles, insofar as the mercury acquires a great subtlety each time. He also compares the earth that the Mercury casts off to the tail left behind by the common mercury when it is not sufficiently purified. He said 'wash your mercury and purify it by salt and vinegar, until it no longer leaves a tail behind it, while flowing on a plane surface'. We will soon know what he

Nous saurons bientôt ce qu'il entend par sel et vinaigre et nous en avons déjà un aperçu.

Lorsqu'on dissout le Mercure dans l'esprit astral, et qu'on a séparé la terre par décantation et lotion, pour n'en rien perdre, on pose la dissolution dans un lieu frais, et il se fait un dépôt de trois sels savoir, l'un cotonneux, qui nage à la superficie et qui est le mercure; le second qui est aiguillé et de nature du Nitre, et qui est entre deux eaux; et le troisiéme[147] qui est un sel fixe et minéral qui se dépose au fond.

Dans l'état où l'on voit ici le Mercure, il tirerait la teinture des végétaux, et en fairait[148] une médecine. Il est médecin lui même,[149] car si on en mettait la valeur d'un grain au pied d'un arbre prèsque[150] mort et qu'on l'arrosat,[151] il reprendrait une nouvelle vigueur; mais ce serait manger son bled[152] en herbe que d'en rester là; il faut poursuivre le travail.

Quand aux deux autres sels, ils se réduisent en mercure semblable au premier, en continuant l'opération. A cet effet, quand les sels ont été séparés, on dissout la seconde espèce dans l'esprit astral pour en arroser le sel fixe, le dissoudre, le faire fermenter et putréfier: et comme il ne serait pas en assez grande abondance pour terminer l'opération, on acheve[153] les imbibitions avec le Mercure dissout,[154] et on procède comme la première fois, par les poids de nature.

Le poids, si on y fait attention, diffère ici du premier, car la terre n'avait besoin que d'être abreuvée; mais ici il faut que le sel soit dissout et fixé jusqu'à ce qu'il ne puisse plus recevoir d'humidité, qu'il fermente, qu'il pourrisse et donne les mêmes résultats que ci-dessus, c'est à dire[155] un Mercure que vous laverez et dont vous séparerez la terre pour la joindre avec la première.

---

147. troisième.
148. ferait.
149. lui-même.
150. presque.
151. l'arrosât.
152. blé.
153. achève.
154. dissous.
155. c'est-à-dire.

means by salt and vinegar, as we have already had a glimpse.

When we dissolve the Mercury in the astral spirit, and when we separate the earth by decantation and lotion[34] so as not to lose anything, we put the dissolution in a cool place. A deposit of three salts occurs, namely: one that is cottony, which swims to the surface and which is the mercury; the second which is needle-like,[35] and of the nature of Nitre, and which lies between two waters; and the third which is a fixed and mineral salt which settles on the bottom.

In the state in which one sees it here, the Mercury could draw the tincture from plants, and make a medicine of it. It is a medicine in and of itself, for if we put as little as a single grain of it at the foot of an almost dead tree and then watered this tree, it would regain new vigour; but that would be like eating wheat seed rather than letting it mature;[36] the work ought to be continued.

As for the two other salts, they are reduced in mercury similar to the first one, by continuing the operation. To this end, when the salts are separated, dissolve the second kind in the astral spirit in order to water the fixed salt with it, to dissolve it, and to make it ferment and putrefy. And since there would be insufficient quantity at the end of the operation, we complete the imbibitions with the dissolved Mercury, and we proceed as before, by the weight of nature.

The weight, if we pay attention to it, differs from the first, because the earth only needed to be watered. But here the salt must be dissolved and fixed until it can no longer receive any more moisture, until it ferments, putrefies, and gives the same results as above, i.e. a Mercury which you wash, and from which you separate the earth to join it with the first.

34. That is, washing.
35. This term refers to the peculiar, needle-like crystalline structure of the nitre salt.
36. This was a saying used by farmers, dating back to the 16th century, which meant: 'spending money that has not yet been received'.

Pour sublimer le Mercure, vous le séparerez en deux, vous dissoudez[156] une moitié par l'esprit astral, et vous ferez[157] par son moyen des ablutions sur la partie fixe, ainsi que je viens de vous enseigner. Vous continuerez vos ablutions jusqu'à dissolution parfaite, et vous laisserez ensuite fermenter et putréfier comme auparavant.

Vous avez ici le mercure du second aigle; si vous allez ainsi jusqu'au septiéme,[158] inclusivement, ce mercure sera très propre à dissoudre l'or, et il le dissoudra sans chaleur ni ébulition,[159] et à la maniere[160] dont la glace fond dans l'eau chaude; vous le conduirez jusqu'au neuvieme[161] inclusivement, et vous lui donnerez toute l'exaltation dont il est susceptible pour pouvoir opérer de plus grandes choses. Mais, je vous préviens que si vous voulez aller plus loin, il dissoudrait jusqu'au[162] silex par le simple contact et vous ne trouveriez plus de vases pour le contenir.

A chaque sublimation ou aigle, vous séparerez la terre noire féculeuse comme la première fois, et vous la joindrez à la première pour en faire l'usage que je vous indiquerai au second travail; car le premier a été employé tout entier à la façon de notre mercure: mais c'est celui qui exige le plus de tems.[163] Il est aussi le plus difficile, c'est pourquoi il est comparé aux travaux d'hercule dont il n'est au surplus que la juste application: et lorsqu'il est terminé, le reste n'est plus regardé que comme un ouvrage de femme et un jeu d'enfant. Il ne s'agit plus en effet que de laver le laiton, ou de faire une impastation, ce qui s'applique fort bien ou aux femmes qui s'occupent de lessive, ou aux enfans[164] qui font des boulettes et des bonshommes d'argile ou de terre détrempée.

156. dissoudrez.
157. ferez.
158. septième.
159. ébullition.
160. manière.
161. neuvième.
162. jusqu'aux.
163. temps.
164. enfants.

To sublimate Mercury, separate it into two, dissolve one half by means of the astral spirit, and use it to make the ablutions on the fixed part, as I have just taught you. Continue your ablutions until completely dissolved, and then let it ferment and putrefy as before.

Here you have the mercury of the second eagle. If you continue in this way until the seventh, inclusively, this mercury will be sufficient to dissolve gold, and it will dissolve it without heat or boiling, like ice melts in hot water. Push it to the ninth, inclusively, and you will give it all the exaltation that it is capable of in order to perform the greatest things. But I warn you: should you wish to go any further, it would dissolve even flint[37] by simple contact, and you would no longer find any vessels to contain it.

With each sublimation or eagle, separate the starchy black earth as was done during the first time, and join it to the first one to make use of it as I will show you in the second work; because the first [part] is entirely about making our mercury and it is this that requires the most time. It is also the most difficult [part]. This is why it is compared to the labours of Hercules, of which it is, moreover, simply the proper application. And when it is over, the rest is seen as woman's work and child's play. Indeed it is no more than polishing brass,[38] or making a paste, which applies very well to women engaged in laundry, or to children making figures and balls of clay or soggy earth.

---

37. French *silex* generally meant flint, and by extension, any hard rock; but more specifically, silex, or silica, may also refer to an extremely hard, crystalline, white or colourless substance: silicon dioxide, SiO_2, the principal constituent of quartz and sand.
38. French *laiton* may be generally translated as 'brass', but more archaically as 'latten', an alloy of copper and zinc resembling brass. Latten, moreover, refers not to any alloy, but to the Latten of the Wise, i.e. their matter, and thus to the well-known alchemical injunction: *dealbate latonem et libros rumpite*, 'whiten latona and destroy your books'.

Lavare et impastare, in hoc consistit magisterium sapientum.

Le tems[165] de cette grande et importante opération est d'environ deux années communes. Et lorsqu'elle est terminée, l'apprentissage de notre maçonnerie, car il n'est que celle-ci de vraie, cet apprentissage finit, il fait place au compagnonage[166] dont les épreuves sont beaucoup moins longues, et moins rudes.

Vous avez enfin entre les mains ce Mercure universel dont les sages ont tant parlé, par son moyen, vous pouvez attaquer la Nature jusqu'au cœur, et extraire les médecines ou teintures des trois Regnes,[167] en leur donnant en même tems[168] une fixité et perfection qu'elles ne pouvaient avoir. Ce Mercure est véritablement la force de toutes forces dont a parlé le savant Hermes Trismégiste, c'est le dragon igné qui détruit toutes choses, l'esprit de vin,[169] ou plutôt l'eau de vie[170] de Raimon Lulle,[171] et le vinaigre du Cosmopolite. Il dissout et fixe en même tems,[172] car il provient de l'union de deux feux en opposition l'un de l'autre, bien qu'ayant une même origine. Le premier est un feu acide et froid, c'est celui qui dissout et produit la fermentation; le second est alkalin[173] et chaud, il produit la putréfaction et fixe le composé. C'est pourquoi B.V. à la fin de ses *douze clefs* vous avertit de bien distinguer le froid d'avec le chaud, dans l'application de vos feux.

---

165. temps.
166. compagnonnage.
167. Règnes.
168. temps.
169. l'esprit-de-vin.
170. l'eau-de-vie.
171. Raymond de Lulle.
172. temps.
173. alcalin.

Lavare et impastare, in hoc consistet magisterium sapientum.[39]

Altogether, the duration of this great and important operation is approximately two years. And when it has been completed, the apprenticeship of our masonry, the only true one, finishes and gives way to the trade-guilds whose tests are much shorter, and less difficult.

You will finally have in your hands this universal Mercury, of which the wise have spoken so much; and by its means, you can penetrate right into the heart of Nature, and extract the medicines or tinctures from the three Kingdoms, while at the same time giving them a fixity and perfection that they would not have otherwise had. This Mercury is truly the strength of all strengths of which the wise Hermes Trismegistus spoke. It is the igneous dragon which destroys all things, the spirit of wine, or rather the brandy, the living water, of Raymond Lull,[40] and the vinegar of the Cosmopolitan. It dissolves and fixes at the same time, for it stems from the union of two fires in opposition to each other, although having the same origin. The first is an acidic and cold fire; it dissolves and produces fermentation. The second is alkaline and hot; it produces putrefaction and fixes the compound. This is why B.V. at the end of his *Twelve Keys*,[41] warns you to carefully distinguish the cold from the hot in the application of your fires.

39. 'To wash and make paste: the magistry of the wise consists in this'. When the matter putrefies in the egg and becomes black, it thickens to the consistency of black pitch; it is then like a paste or like mud, which is why that operation is called making paste or *impastation*.
40. RAMON(D) LULL(Y), c. 1232–c. 1315, philosopher and logician from the former Kingdom of Majorca (now part of Spain). Although many alchemical texts are attributed to him, it is probable that some were composed by some of his followers or disciples.
41. BASIL VALENTINE's *Twelve Keys* (*Zwölf Schlussel*) first appeared in *Ein kurtz summarischer Tractat, von dem grossen Stein der Uralten* (1599), with a French translation appearing in 1678.

Ce n'est pas pourtant que la chaleur fermentative provienne de l'alkali[174] plutôt que de l'acide, puisqu'elle n'est qu'un simple effet du mouvement, comme vous avez du[175] le remarquer au commencement de ce traité; mais parce que la présence de cet alkali[176] la détermine et la conserve pendant la putréfaction.

Le Mercure n'étant qu'une demi génération,[177] il faut procéder maintenant à l'exaltation du Soufre. Ainsi que l'ont fait Flamel et Le Trévisan, vous pouvez prendre de l'or en feuilles et en extraire la teinture en la projetant dans votre Mercure que vous aurez dissout auparavant. Cette voie n'est pas la plus noble, mais elle est la plus courte; ce n'est qu'une teinture particulière qu'on obtient, mais le Mercure l'universalise dans le travail et la conduit au même résultat.

Il est bien plus noble sans doute de tirer de la matière cette teinture universelle. Vous prendrez donc toutes vos terres provenant des aigles, et vous procéderez avec elles par de nouvelles imbibitions avec l'esprit astral, jusqu'à ce qu'elles rougissent et qu'elles soient d'un rouge-brun. C'est ce que les pphes[178] appellent la calcination. Le Mercure dissous et projeté dessus faira[179] l'extraction de la Teinture, au moyen de laquelle vous pourrez procéder au Mariage Philosophique qui fera la perfection de l'œuvre, et terminera les travaux, sauf la multiplication qui n'en est que la répétition abrégée.

Cette Teinture est la couronne du Roi que vous devez tirer des cendres, pourquoi le sage pythagore et après lui plusieurs ont répété, Ne méprisez pas les cendres, parce que

---

174. alcali.
175. dû.
176. alcali.
177. demi-génération.
178. philosophes.
179. fera.

It is not however that the fermentative heat comes from the alkali rather than from the acid, since it is only a simple effect of a process, as you must have noticed at the beginning of this treatise, but because the presence of this alkali determines it and preserves it during the putrefaction.

Because Mercury represents only half of the generation, we must now proceed to the exaltation of Sulphur. As did Flamel and Trevisan, you can take gold leaf and extract the tincture from it by projecting it into your Mercury which you have previously dissolved. This way is not the noblest, but it is the shortest. Only a particular tincture is obtained, but the Mercury universalises it in the work and leads to the same result.

Without doubt, it is much nobler to draw this universal tincture from the matter. Thus you will take all your earths resulting from the eagles, and proceed with them to new imbibitions with the astral spirit until they redden and then turn reddish-brown. This is what the philosophers call the calcination. The Mercury, dissolved and projected on top of it, will extract the Tincture, by means of which you can proceed to the Philosophical Marriage, which will be the perfection of the work, completing the labours, save for the multiplication which is only a short repetition.

This Tincture is the crown of the King that you must extract from the ashes, which is why the sage Pythagoras and many after him have repeated: 'Do not despise the ashes, for the crown of the King is contained therein'.[42] From

---

42. PYTHAGORAS is presented as a Hermetic adept in the *Turba Philosophorum*, as well as in Arabic alchemical tradition. Although the *Greek Alchemical Corpus* does not mention Pythagoras, the pivotal significance of the ashes (τεφραι) is attributed to 'all the [ancient] philosophers' by Olympiodorus (BERTHELOT, CAAG, 1.3.4, 37). Cf. especially Peter KINGSLEY's important study, 'From Pythagoras to the *Turba Philosophorum*: Egypt and Pythagorean Tradition', *Journal of the Warburg & Courtauld Institutes*, 1994, vol. 57, pp. 1-13.

la couronne du Roi s'y trouve renfermée. C'est de là que provient la coutume de conserver la cendre des morts. B.V. dit dans sa préface « que la couronne du Roi soit de très pur or; et ailleurs il dit: *C'est une couronne tirée des cendres*. L'or est cette Teinture dont nous parlons, et la cendre est la terre des aigles que vous avez mise[180] à part.

Il faut aussi que vous sachiez que le Mercure, qui fait l'extraction de cette Teinture, est appelé Eau sèche qui ne mouille pas les mains, parce que, bien qu'il ne soit qu'un sel qui ne mouille point, il a seul la vertu de dissoudre tous les corps, ainsi que l'eau fait des sels et des Gommes. En apparence, l'eau est dite un dissolvant, mais, au fait, elle ne fait que diviser. La dissolution n'a lieu dans toute la nature qu'au moyen de la fermentation, tandis que le Mercure en dispense dans les mêmes occasions; mais dans les choses plus élevées où la présence de l'eau est de nul effet, et en remplit les fonctions, et ne fait comme elle, que séparer les corps ou substances pour les mettre aux prises, et leur faire subir la fermentation, seule cause de dissolution. Au surplus la dissolution n'est elle même[181] qu'une division plus étendue des corps, ou une disjonction absolue, et le mêlange[182] exact de toutes leurs parties. Il arrive en cette circonstance que les parties disjointes et d'une nature opposée entre elles venant à se rencontrer, se heurtent et se livrent une espèce de combat auquel nous avons donné le nom de fermentation, après quoi elles s'unissent de nouveau, mais après s'être purgées de ce qui leur était étranger qui cause la corruption, et empêche que l'union ne soit parfaite; mais après son entière séparation, l'union est si intime que tous les efforts de la Nature pour les séparer seraient nuls et insuffisants. Ainsi seront les corps et les ames[183] des justes après le jugement et leur purification.

---

180. mise.
181. elle-même.
182. mélange.
183. âmes.

this comes the custom of preserving the ashes of the dead. B.V. said in his preface that 'the crown of the King will be of very pure gold'; elsewhere he says: '*It is a crown drawn from the ashes*'.[43] The gold is this tincture of which we speak, and the ash is the earth of the eagles that you have set aside.

You must also know that Mercury, which causes the extraction of this Tincture, is called 'dry Water which does not wet the hands', because, although it is only a salt which does not moisten, it alone has the virtue of dissolving all bodies, just as water does with salts and Gums. Apparently, water is known as a solvent, but in reality it only divides. Dissolution takes place in all of nature only by means of fermentation. Mercury provides the same opportunities for dissolution in more elevated things[44] upon which the presence of water has no effect. Here Mercury performs the same functions as water, but does so differently; it separates the bodies or substances in order to put them into conflict, and makes them undergo fermentation, which is the sole cause of dissolution. Furthermore, dissolution itself is only a more thorough division of bodies, an absolute separation, and the perfect mixing of all their parts. In this circumstance, what happens is that the separated parts, which are of opposed natures, suddenly encounter each other, collide, and engage in a type of combat to which we have given the name of 'fermentation'. After this they are reunited, but only after having purged themselves of that which was foreign to them, that which causes corruption and prevents the union from being perfect. But after its absolute separation, the union is so intimate that every effort of Nature to separate them again would be wholly insufficient and amount to nothing. And so too will the bodies and souls of the righteous be after the judgment and their purification.

---

43. VALENTIN(US), *Les Dovze clefs de philosophie*, Elise Vveyerstraten, Amsterdam, 1678, p. 4.
44. That is, things that are difficult to dissolve.

Après l'extraction de la Teinture, il reste en arrière une terre réfractaire que nous appelons terre damnée, parce que, comme le péché, elle est cause de mort et de souffrances. Il faut la rejeter avec soin, car c'est elle qui empêche l'ingrès de la teinture, et qui cause ici bas[184] lanthipatie[185] et l'inimitié parmi les êtres.

L'ébullition qui accompagne ordinairement la fermentation est figurée dans nos livres comme un combat entre deux champions dont l'un doit surmonter l'autre, et le mettre à mort; mais il ne faut pas prendre tout à fait à la lettre. Cette ébullition ne doit être attribuée qu'au dégagement des Gaz qui cherchent à se mettre en équilibre, soit par mixtion, soit par extension.

De même, lorsque nous parlons de Sceau Hermètique[186]; il ne faut pas l'entendre de la cloture[187] exacte du vase: cloture[188] imbécile et qui serait plus nuisible qu'utile, attendu qu'elle empêcherait la manipulation aussi bien que la séparation et conjonction des substances dans les temps et proportions dues. Nous appelons ainsi la réunion de plusieurs substances en une seule, de manière à ne pouvoir plus les séparer: car chez nous, ou dans notre langage, ouvrir est la même chose que dissoudre, et fermer, la même chose que fixer. Nous avons sept sceaux correspondant à sept corps planétaires, et qui connait[189] l'un, connait[190] tous les autres.

Nous nous servons aussi de beaucoup de termes familiers à la chimie vulgaire; il faut que l'on sache, une fois pour toutes, que distiller, cohober, sublimer, calciner, réverbérer, incérer &c. ne sont chez nous depuis le commencement jusqu'à la fin, qu'une seule et même opération, laquelle consiste à dissoudre et coaguler, ce qui est la même chose que mouiller et dessécher, et que le moindre apprentif sait faire.

Maintenant que vous avez la solution des Enigmes principales qui obscurcissent notre langage et en empêchent

---

184. ici-bas.
185. l'antipathie.
186. Hermétique.

187. clôture.
188. clôture.
189. connaît.

190. connaît.

After the extraction of the Tincture, a refractory earth remains behind which we call 'damned earth', because, like sin, it is the cause of death and suffering. It should be rejected with diligence, for it prevents the ingress of the tincture, and among beings here below, causes antipathy and enmity.

The boiling or ebullition which usually accompanies fermentation is depicted in our books as a combat between two champions in which one must overcome the other, and put him to death. Yet it must not be taken altogether literally. This 'boiling' should only be attributed to the release of gases which seek equilibrium, either by mixture, or by expansion.

It is the same when we speak of the Hermetic Seal. It should not be understood as the exact closure of the vessel, which would be a foolish closure: more harmful than useful, since it would prevent manipulation as well as the separation and conjunction of the substances in due times and proportions. Rather, we use it to refer to the reunion of several substances into a single substance, such that they can no longer be separated, because to us, or in our language, to open is the same as to dissolve, and to close is the same as to fix. We have seven seals corresponding to seven planetary bodies, and whosoever knows the one, knows all the others.

We also use many other terms familiar to vulgar chemistry. It must be known, once and for all, that to distil, to cohobate, to sublimate, to calcine, to reverberate, to incinerate etc., are for us, from the beginning to the end, but one and the same operation, which consists in dissolving and coagulating, which is the same thing as to moisten and to desiccate, which any apprentice should be able to do.

Now that you have the solution to the principal Enigmas which obscure our language, and which prevent, or at least delay the understanding of it, I am going to explain the

ou retardent au moins l'intelligence, je vais vous expliquer ce que c'est que notre mariage pphique[191] entre *Baya*[192] et *Gabertin*. Vous devez savoir à présent que la Teinture rouge, qui est le Soufre fixe des pphes,[193] et qu'ils appellent tantôt Lion, tantôt esprit de vin[194] ou vinaigre très aigre, et quelquefois orpiment fait ici fonction de mâle et est appelé Gabertin. Le Mercure ou la Teinture blanche qu'ils nomment Lune, argent, Eau de vie , vinaigre, arsenic, magnésie, Terre feuillée &c. fait ici l'office de femelle et est appelée Beya.

Il faut savoir encore que ces deux substances, soufre et Mercure que le *petit paysan* appelle les deux fleurs, ne constituent ensemble qu'un seul Mercure, dit hermaphrodite, ou plutôt androgine,[195] qui signifie mâle et femelle ; que dans l'opération que je vais décrire, elles en font alternativement les fonctions; que par conséquent ils ont souvent donné à l'un et à l'autre les mêmes noms, mais particulièrement celui de Mercure, en faisant pourtant une petite différence essentielle à connaitre[196]; ils mettent alors devant le nom de Mercure le mot premier, pour exprimer la teinture blanche. Ils nomment celleci[197] Lion vert, et le Soufre Lion Rouge. S'ils nomment le Mercure eau de vie,[198] vinaigre,[199] arsenic, magnesie,[200] Lune, argent, ils nomment par une juste comparaison et proportion la Teinture rouge, Esprit de vin,[201] le vinaigre très aigre, orpiment, réalgar, or vif, Soleil &c.

191. philosophique.
192. Beya.
193. philosophes.
194. esprit-de-vin.
195. androgyne.
196. connaître.
197. celle-ci
198. eau-de-vie.
199. vinaigré.
200. magnésie.
201. Esprit-de-vin.

meaning of our philosophical marriage between *Beya* and *Gabritius*.[45] You must know by now that the Red Tincture, which is the Fixed Sulphur of the philosophers, and which they sometimes call Lion, sometimes spirit of wine or acrid vinegar, and sometimes orpiment, here plays the role of male, and is called Gabritius. Mercury or the White Tincture, which they name Luna, silver, Living Water (Brandy), vinegar, arsenic, magnesia, Foliated Earth, etc., here acts as female and is called Beya.

It should also be known that these two substances, sulphur and Mercury, which the author of *Le Petit Paysan*[46] calls the two flowers, together constitute a single Mercury, known as the hermaphrodite, or rather the androgyne, which means both male and female. In the operation that I will describe, their functions alternate. Consequently they have often given the same names to one and the other, but particularly that of Mercury. However there is a small difference which is essential to know. They put before the name of Mercury the word 'first', to designate the white tincture. They name the latter the Green Lion, and Sulphur the Red Lion. If they call Mercury Living Water [Brandy], vinegar, arsenic, magnesia, Luna, and silver, then by a fair comparison and proportion they call the Red Tincture Spirit of Wine, acrid vinegar, orpiment, realgar, living gold, Sun, etc.

---

45. Usually presented as sister and brother, wife and husband, Gabritius (*Gabertin* in the manuscript) derives from Arabic *kibrīt*, 'sulphur', while Beya appears to come from *al-baiḍā*, 'white, bright', referring to mercury. Cf. Julius RUSKA, *Turba Philosophorum: Ein Beitrag zur Geschichte der Alchemie*, Berlin, Julius Springer, 1931, p. 324.
46. Known in German as *Der grosse und der kleine Bauer*, the two treatises were first published in *Aperta Arca arcani artificiosissimi* (1617), attributed by some to the Pomerian jurist Johann GRASSHOFF (1560–1623). The edition that our author refers to, however, is the French edition, *Le Petit Paysan*, Strasbourg, 1619; cf. Nicolas LENGLET DUFRESNOY, *Histoire de la Philosophie Hermetique,* Paris, Coustelier, 1742, tome III, p. 259.

Pour dernière observation, je vous fairai[202] remarquer que le mercure n'est qu'un sel inverti en cette substance mercurielle ; que le Souffre[203] lui même[204] n'est jamais sans Sel, non plus que le Sel sans Mercure : ce qui vous fait voir jusqu'à l'évidence trois substances en une, lesquelles substances nous appelons, pour notre commodité, Sel, Soufre et Mercure.

Pour procéder au mariage pphique,[205] vous séparez en deux votre Teinture Rouge, et vous en laissez dessécher une partie, mettant l'autre à part pour le besoin. Combien de gens ont failli, pour avoir ignoré cette précaution ! Ils ont cru que blanchir le rouge, et rougir le blanc, n'était qu'une suite ordinaire et nécessaire de la marche du grand Œuvre, et que tout cela se faisait de soi même.[206] Qu'ils sachent donc que le rouge est nourri du blanc et le blanc du rouge ; que le blanc est pris pour le lait dont on nourrit l'enfant nouveau né,[207] ou pour la Robe virginale. Quant au rouge, il exprime ou l'augmentation du feu, ou le changement de vêtement, il est pris par quelques uns[208] pour le Manteau Royal.

Vous procéderez donc aux imbibitions sur une moitié de votre Soufre, que vous aurez laissé dessécher, avec le Mercure blanc, suivant les poids et mesures dont vous avez déja[209] fait usage, et continuerez ainsi jusqu'à une complette[210] saturation et que la matière demeure liquide au fond du vaisseau, c'est à dire[211] boueuse. Si vous avez bien opéré, vous obtiendrez en quarante jours la dissolution du corps, à la suite de laquelle viendront la fermentation et la putréfaction.

Dans la fermentation, la matière se boursoufle, s'élève et fait un petit bruit comme celui d'une fourmilière; et lor-

---

202. ferai.
203. soufre.
204. lui-même.
205. philosophique.
206. soi-même.
207. nouveau-né.
208. quelques-uns.
209. déjà.
210. complète.
211. c'est-à-dire.

As a final observation, I will point out that mercury is only a salt inverted into this mercurial substance, that Sulphur itself is never without Salt, nor is Salt ever without Mercury: which reveals the obvious reality of three substances in one. For our convenience, we call these substances Salt, Sulphur, and Mercury.

To proceed to the philosophical marriage, separate your Red Tincture into two, desiccate one part of it, and set the other aside for later. How many people have failed because they ignored this precaution! They believed that to whiten the red, and to redden the white was only a common and necessary consequence in the progress of the great Work, and that all this was done by itself. They should instead know that the red is nourished by the white and the white by the red, and that the white is taken for the milk with which one feeds the newborn child, or for the virginal Robe. As for the red, it expresses either the augmentation of the fire, or the change in vestments. It is taken by some as the Royal Mantle.

You shall thus proceed to the imbibitions on the half of the Sulphur that you allowed to desiccate, using the white Mercury, according to the weights and measures of which you have already made use. Continue as such until a complete saturation is achieved, and the matter remains fluid at the bottom of the vessel, i.e. muddy. If you have operated well, then in forty days you will obtain the dissolution of the body, after which fermentation and putrefaction will follow.

During fermentation, the matter swells, rises, and makes a noise like a hive of ants. And when the putrefaction is ready to take place, the matter subsides and black-

sque la putréfaction veut arriver, la matière s'affaisse et noircit. Ce n'est que lorsqu'elle est arrivée à la noirceur parfaite, nommée tête du Corbeau, qu'elle est en pleine putréfaction. C'est là seulement la première matière de notre Œuvre, matière qu'on ne trouve nulle part sur la terre des vivants, qu'on ne crée pas cependant, mais qui est dite avoir *volée*[212] au-dessus[213] de nos têtes, à cause que le mercure ayant été sublimé neuf fois, le Soufre s'est encore élevé par dessus.[214]

Les pphes[215] prennent la dissolution pour le regne[216] de Mercure; c'est pendant ce Regne[217] que s'allient entre eux nos principes métalliques, mais il est ici comme hors d'œuvre[218]; ce n'est ici qu'un Regne[219] de Saturne ou pendant la noirceur qu'ils commencent à compter, ou qu'ils prennent le commencement de l'œuvre, parce que les trois principes sont liés d'une manière irrévocable et que le Sceau d'Hermès est accompli. C'est le vase de Nature qu'il faut fermer et non un œuf de cristal ou de tout autre matière ; et la cloture[220] ne s'entend pas de la gorge d'un vase pour que l'air n'y puisse pénétrer, mais de la jonction intime du sel et du Soufre et du Mercure, de manière à ce qu'on[221] ne puisse plus les séparer par tel art que ce soit.

Il n'y a besoin d'aucun feu externe pour arriver à la blancheur, la matière en se desséchant y arrive d'elle même.[222] Dabord,[223] elle prend la couleur de gris cendré que l'on compare à l'Etain,[224] et que l'on appele[225] le sceau de Jupiter ; ensuite elle arrive par degrès[226] à la blancheur ; mais avant d'y arriver, on apperçoit[227] circulairement sur la matière diverses couleurs, rouges, jaunes, bleues et vertes que l'on compare à l'iris ou arc-en ciel,[228] et que d'autres

212. volé.
213. au-dessus.
214. par-dessus.
215. philosophes.
216. règne
217. Règne.
218. hors-d'œuvre.
219. Règne.
220. clôture.
221. que l'on.
222. d'elle-même.
223. D'abord.
224. l'Étain
225. appelle.
226. degrés.
227. aperçoit.
228. arc-en-ciel.

ens. Only when it has reached the perfect blackness known as the 'Raven's head' is it in full putrefaction. This is only the first matter of our Work, a matter that is nowhere to be found in the land of the living, a matter which has not been created, but which is said to have *flown* over our heads, because after the mercury has been sublimated nine times, Sulphur still rises above it.

The philosophers assign dissolution to the reign of Mercury. It is during this Reign that our metallic principles ally themselves with one other, but this is akin to an appetiser. It is the Reign of Saturn, during the blackness, that matters to them, and which they take as the beginning of the work, because here the three principles are linked in an irrevocable way, and the Seal of Hermes is accomplished. It is the vase of Nature which should be closed, and not a crystal egg or any other such matter. And the closure does not mean that the neck of a vase is sealed so that air cannot penetrate, but rather the intimate juncture of Salt, Sulphur, and Mercury such that no art whatsoever can separate them.

There is no need for external fire to reach the whiteness. The matter, by desiccating, arrives there by itself. Initially, it takes on an ash grey colour, which is compared to Tin, and which is called the 'seal of Jupiter'. Then it gradually reaches whiteness. But beforehand, we can see various colours circulating upon the matter—reds, yellows, blues, and greens—which we compare to the iris or rainbow, and which others call the 'tail of the Peacock'. These colours, which do not last long, are replaced by a film of blackish brown which is striated by desiccation, allowing us to see

appelent[229] la queue du Paon. Ces couleurs, qui ne durent guéres,[230] sont remplacées par une pellicule d'un brun noiratre[231] qui se strie par dessiccation et laisse voir la matière sous une couleur grise : bientôt après on apperçoit[232] sur les bords du vase un cercle capillaire d'une grande blancheur; alors, le Regne[233] de Jupiter, qu'annonçait la couleur grise, et que les pphes[234] comparent au feu de cendres, finit, pour faire place à celui de la Lune. Ce Cercle s'agrandit successivement jusqu'à la blancheur parfaite de la matière que les pphes[235] appelent[236] avec raison Lune ou Argent, puisqu'un poids de cette médecine blanche projeté sur 10 d'argent, et ensuite sur 100 d'un autre métal imparfait, transmue celui-ci en argent plus pur que celui des Mines.

L'argent que l'on emploit[237] en cette circonstance, tient ici lieu de ferment, et sans lui il n'y aurait pas de transmutation, c'est dans ce sens qu'il faut entendre ce que disent les sages: que sans or, aucun or n'est faisable; ils entendent parler du ferment.

Cette terre blanchie a l'aspect d'une poudre brillante de diamant, et est divisée en petites lames : ce qui est cause que les sages l'ont nommée leur terre feuillée dans laquelle ils recommandent de semer leur Or, elle n'est comme l'on voit qu'une demi génération,[238] c'est pourquoi il faut continuer le travail si l'on veut arriver à la perfection.

Il faut donner à cette terre la culture nécessaire avant d'y semer l'or, autrement il ne fructifierait point.

On recommence donc les Imbibitions avec le mercure blanc, selon la mesure antérieurement observée. A l'aide d'un feu bien observé, la matière se subtilise de plus en plus, se couvre de verdure, après quoi elle commence à jaunir et prend une couleur orangée qu'elle ne pourrait plus dépasser si le feu n'était augmenté.

229. appellent.
230. guère.
231. noirâtre.
232. aperçoit.
233. Règne.
234. philosophes.
235. philosophes.
236. appellent.
237. emploie.
238. demi-génération.

the grey-coloured matter beneath. Soon afterwards, we will perceive a capillary circle[47] of a great whiteness on the edges of the vase. Then the 'Reign of Jupiter', which announces the grey colour, and which the philosophers compare to the fire of ashes, is complete, giving way to the 'Reign of the Moon'. This Circle expands successively until the matter reaches perfect whiteness, which the philosophers, with good reason, call the Moon or Silver, because a measure of this white medicine, projected onto ten measures of silver, and then onto a hundred measures of another imperfect metal, transmutes them into a silver purer than that of the Mines.

The silver which we use in this circumstance takes the role of a ferment or leaven, and without it there would be no transmutation. It is in this sense that we should understand the Sages, for when they say: 'without gold, no gold is possible', they are speaking of the ferment.

This whitened earth has the appearance of a brilliant diamond powder, and is separated into small flakes or lamellæ.[48] This is why the wise have named it their 'foliated earth' and advise us to sow their Gold in it. But this is only viewed as a half-generation, which is why it is necessary to continue the work if we want to achieve perfection.

It is necessary to cultivate this earth before sowing the gold therein, otherwise it would not bear fruit.

We therefore recommence the imbibitions with the white mercury, according to the measurement previously observed. With the aid of a well-observed fire, the matter becomes increasingly subtle, covers itself with greenness, after which it begins to yellow, taking on an orange colour that cannot be exceeded unless the fire is increased.

---

47. Literally 'hairlike circle' or 'hairlike ring' (*cercle capillaire*). See discussion under note 33.
48. French *lames*, from Latin *lamellæ*, referring to small, thin plates of of metal; metallic leaf or foil.

Cette verdure tant chantée par les poëtes,[239] et si recommandée par tous les pphes[240] est le regne[241] de la belle venus,[242] auquel succède celui de Mars qui est la couleur orangée.

Vous vous souvenez d'avoir fait deux parts de votre teinture Rouge : vous venez de blanchir la première, il faut maintenant la rougir. Prenez donc la Teinture mise en réserve, dissolvez-la en projetant dessus du mercure pphique[243] et procédez avec cette Teinture aux imbibitions, jusqu'à ce que la matière arrive à un beau rouge pourpré et foncé de pavot.

Telle est la médecine du premier ordre, tant au Blanc qu'au Rouge, laquelle guérit toutes maladies lorsqu'on en use sans addition de métal, dans un véhicule approprié au mal, selon la prudence requise, et qui avec l'addition, comme ferment, des deux métaux parfaits, transmue en or ou en argent tous les métaux imparfaits, tels que le cuivre, le plomb, l'étain &c.

Auparavant que de tenter une projection, il faut essayer la matière sur une lame de Cuivre rougie au feu. Si elle fond sans fumée elle est dans l'état désiré, autrement il faudrait continuer le feu.

Multiplication

La Multiplication n'est autre chose que la répétition de tout l'Œuvre, à partir du mariage philosophique. Il faut seulement avoir le soin de partager en deux sa matière dans le Cercle de la blancheur et dans celui de la rougeur, afin de pouvoir procéder aux imbibitions sur la moitié restante avec

---

239. poëtes.
240. philosophes.
241. règne.
242. Vénus.
243. philosophique.

This greenness, of which the poets sing, and which all the philosophers recommend, is the Reign of beauteous Venus, which succeeds the orange of the Reign of Mars.

You recall having divided your Red Tincture into two parts: you have just whitened the first, it now needs to be reddened. Thus, take the Tincture that was set aside, dissolve it by projecting some philosophical mercury onto it, and proceed with this Tincture to the imbibitions until the matter reaches the beautiful, dark purple-red of the poppy.

Such is the medicine of the first order, both the White and the Red, which cures all diseases when we use it without the addition of metal, in a vehicle appropriate to the illness, according to required prudence; a medicine which, with the addition, as ferment, of two perfect metals, transmutes all imperfect metals such as copper, lead, tin, etc., into gold or silver.

Before attempting a projection, the matter should be tested on a piece of red hot Copper foil. If it melts without smoke, it is in the desired state. Otherwise it is necessary to continue the fire.

Multiplication

Multiplication is nothing but the repetition of the whole Work, starting from the philosophical marriage. It is only necessary to carefully divide the matter into two parts during the Cycle of whiteness and during that of redness, in order to be able to proceed to the imbibitions on the remaining half with parents of the same blood. Here, the Mercury and the Red Tincture, in their first State,

des parens[244] d'un même sang. Le Mercure aussi bien que la teinture Rouge dans leur premier Etat,[245] seraient ici trop imparfaits pour pouvoir s'allier à notre médecine.

Vous aurez soin à chaque dissolution par le mercure de séparer une terre damnée qui se précipite et que vous rejeterez[246] avec d'autant moins de scrupule, qu'elle est absolument réfractaire, et qu'elle empêche l'ingrès de la matière dans les métaux.

Avec toutes les conditions que j'ai décrites ci-dessus, sans en rien omettre, vous arriverez surement[247] au but si désiré de la Philosophie.

Toutes fois,[248] ne cherchez pas à outrepasser le nombre sacré de neuf, car la matière, si fixe qu'elle soit, aurait acquis une si grande fluidité et dilation, qu'aucun vase ne pouvant la contenir, elle serait entièrement perdue.

Sur ce, mon frere,[249] remerciez Dieu de la grace[250] qu'il vous a faite, ainsi que je le remercie de vous avoir été utile dans vos desseins, s'ils sont droits, et que vous demeuriez dans les sentiers du bien.

fin

---

244. parents.
245. État.
246. ejetterez.
247. sûrement.
248. Toutefois.
249. frère.
250. grâce.

would be too imperfect to be able to ally themselves with our medicine.

You should take care, with each dissolution by Mercury, to separate the condemned earth which precipitates, and which you will reject without the least scruple. It is absolutely refractory, and prevents the ingress of the matter into metals.

With all the conditions that I have described above, without omitting anything, you will surely arrive at the desired goal of Philosophy.

However, do not try to exceed the sacred number nine, for however fixed the matter may be, it will have acquired so much fluidity and dilation that no vessel could contain it, and it would be lost entirely.

On that account, my brother, thank God for the grace he has given you, as I thank him for having served your purposes, if your motives are upright, and you remain on the proper pathways.

end

Scholies
(commentaire)

1er

Tout était eau dès le principe: l'Univers et tout ce qu'il renferme est sorti des Eaux.

2e

L'Eau est un composée de divers principes, si cela n'était pas, elle n'éprouverait pas de fermentation ni de putréfaction.

3e

L'Eau fermentée, pourrie et desséchée forme un limon que l'on peut appeler Eau sèche.

4e

Ce Limon, cette Eau sèche, c'est l'argile dont le Colosse du monde a été formé.

5e

L'Argile est une Terre onctueuse, grise et pesante dont on fait la Brique.

Scholium
(Commentary)

1st

'Everything is water' derives from the principle: the Universe and all that it contains comes from the Waters.

2nd

Water is a composition of various principles. If it was not, it could not undergo fermentation or putrefaction.

3rd

Water fermented, putrefied, and desiccated forms a silt that can be called Dry Water.

4th

This Silt, this Dry Water, is the clay from which the Colossus of the world was formed.

5th

Clay is an unctuous, heavy, grey Earth from which one makes Brick.

6E

L'alcalescence et non la graisse forme son onctuosité, et la rend savonneuse.

7E

C'est ce qui la rend miscible avec les corps gras, mais non d'une façon intime: à la moindre chaleur, la graisse se sépare.

8E

L'Argile n'est donc pas formellement un Alkali[1] ; mais il a une qualité voisine de sa nature. Il tient l'intermédiaire.

9E

Il passe souvent à l'état de craie ou de chaux, mais imparfaitement, il conserve en plus ou moins grande partie sa forme première.

10E

Les terres jaunes, rouges, vertes, &c. sont de cette Nature, mais avec addition de Teinture Minérale.

11E

Cette Teinture est produite par mutation, d'une partie de la terre première en vitriol de la nature du fer ou du Cuivre.

---

1. Alcali.

6TH

Alkalescence and not fat forms its unctuousness, and renders it soapy.

7TH

This is what makes it miscible with fatty substances, but not in an intimate way: with the slightest heat, the fat separates.

8TH

Clay is therefore not technically an Alkali; but it has a quality close to its nature. It is an intermediary.

9TH

It often passes to the lime or chalk state, but imperfectly, it retains more or less the greater part of its initial form.

10TH

The earths which are yellow, red, green, etc., are of this Nature, but with the addition of the Mineral Tincture.

11TH

This Tincture is produced by mutation of a part of the primary earth into the vitriolic of the nature of iron or copper.

12E

La double action de l'Esprit aërien[2] et de l'esprit minéral, opérent[3] ces diverses mutations.

13E

L'Esprit Astral, aërien[4] et universel introduit dans ce sujet, suivant sa pureté, lui donne une forme plus ou moins noble.

14E

La pierre, le Marbre, les sels, les Cristaux et les Minéraux tirent leur origine de cette Terre.

15E

L'Argile est la matrice naturelle et première du monde entier: l'Esprit astral en est la semence.

16E

L'Esprit astral est sans équivoque la lumière du soleil et des astres dont l'air et les cieux sont remplis.

17E

Dans notre système terrestre, le soleil est le père de cet esprit, la Lune en est la mère.

18E

La Lune est dite la mère de l'Esprit astral, parce que sa Lumière vivifique tire sa source du soleil.

2. aérien.
3. opèrent.
4. aérien.

12TH

The double action of the Ærial Spirit and the mineral spirit, operate these diverse mutations.

13TH

The Astral Spirit, ærian and universal, introduced into this subject, according to its purity, gives it a more or less noble form.

14TH

Stone, Marble, salts, Crystals, and Minerals draw their origin from this Earth.

15TH

Clay is the first, natural matrix of the entire world: the Astral Spirit that it contains is the seed.

16TH

The Astral Spirit is unequivocally the light of the Sun and the stars, with which the air and the heavens are filled.

17TH

In our terrestrial system, the Sun is the father of this spirit, the Moon is the mother.

18TH

The Moon is said to be the mother of the Astral Spirit, because its vivifying Light draws its source from the Sun.

19E

Cependant tous les astres y joignant leur lumière, son véritable nom est l'Esprit universel.

20E

Il faut que cet esprit qui est un feu, soit dissout par un autre feu, et devienne Eau.

21E

On recuille[5] cet Esprit dans la grande mer des sages qui est l'air, par le moyen d'un acier magique qui est d'une même nature.

22E

Le feu central renfermé dans tous les corps est un acier magique.

23E

Ce mot magique vous fait voir que ce n'est point un véritable acier, mais qu'on ne l'appelle[6] ainsi que par comparaison.

24E

Tous les corps qui ont vie attirent l'air pour leur nourriture. Le regne[7] animal est celui où cette attraction se fait le plus visiblement.

---

5. recueille.
6. l'appele.
7. règne.

19TH

However when all the stars combine their light, their true name is Universal Spirit.

20TH

It is necessary for this Spirit, which is a fire, to be dissolved by another fire, and become Water.

21ST

We collect this Spirit in the great sea of the wise which is the air, by means of a magical steel which is of the same nature.

22ND

The central fire enclosed in all bodies is a magical steel.

23RD

This word 'magic' enables you to see that it is not truly a steel, but that it is only called so by comparison.

24TH

All bodies that have life draw air for their nourishment. The animal kingdom is the one where this attraction occurs most visibly.

25e

Aussitôt que l'esprit astral est attiré, il est réduit en eau dont les sages font leur feu secret.

26e

Quoique tous les tems[8] soient propres à cette attraction, le printems[9] est la saison la plus convenable, ensuite l'automne.

27e

A ces deux époques, la Nature se régénère, et l'air est plus chargé de cet esprit vital.

28e

La Lune étant la mère de cet esprit, ce n'est que quand elle luit qu'elle nous le donne.

29e

Par conséquent, plus sa lumière est grande, plus cet esprit est abondant.

30e

La Terre est ronde, et son mouvement est d'occident en orient.

31e

L'esprit repoussé vers les Pôles par ce mouvement, et ne trouvant son repos que vers le Nord, il s'y réfugie.

---

8. temps.
9. printemps.

25TH

As soon as the astral spirit is drawn, it is reduced to water from which the wise make their secret fire.

26TH

Though all times of the year are considered apropriate for this attraction, spring is the most suitable season, then autumn.

27TH

At these two times of the year, Nature regenerates herself, and the air is more charged with this vital spirit.

28TH

The Moon, being the mother of this spirit, only gives it to us when she shines.

29TH

Consequently, the greater its light, the more abundant the spirit.

30TH

The Earth is round, and its movement is from the west to the east.

31ST

The spirit is repelled towards the Poles by this movement, only finding its rest towards the North, where it takes refuge.

32e

Le Nord étant sa patrie, c'est dans cette région de l'athmosphère[10] qu'on doit en faire la récolte.

33e

Aussitôt que le soleil paraît sur l'horizon, il chasse l'esprit, il faut cesser le travail.

34e

Esaü[11] vendit à Jacob son droit d'aînesse pour un plat de lantilles,[12] il faut diviser ainsi sa terre.

35e

Il faut faire pleuvoir sur cette terre la rosée du ciel, c'est-à-dire, l'esprit, et qu'elle en soit imbibée.

36e

Que la terre ne soit ni trop abreuvée, ni pas assez, mais qu'elle demeure mouillée.

37e

Ce que la terre peut contenir d'humidité, est le poids de nature. La terre qui contient est le vase.

38e

Il ne faut rendre l'eau à la terre qu'après sa parfaite dessication.[13]

---

10. l'atmosphère.
11. Esaû.
12. lentilles.
13. dessiccation.

32ND

North being its homeland, it is in this region of the atmosphere that it must be harvested.

33RD

As soon as the Sun appears on the horizon, it drives out the spirit, and it is necessary to cease the work.

34TH

Esau sold Jacob his birthright for a dish of lentils; it is necessary to divide the earth in this way.

35TH

It is necessary to make the heavenly dew, i.e. the spirit, rain on this earth, so that the earth is imbibed.[1]

36TH

The earth should neither be watered too much, nor too little, but enough so that it remains moist.

37TH

The amount of moisture that the earth can contain is the weight of nature. The earth which contains it is the vessel.

38TH

It is necessary to return the water to the earth only after its perfect desiccation.

---

1. That is, 'soaked', 'absorbed'.

39E

Mouiller et dessécher, composent le jour naturel.

40E

Chaque humectation est appelée cohobation, et chaque dessication[14] distillation.

41E

A chaque imbibition, le feu central retient du feu Secret la portion spirituelle, le flegme se dissipe entièrement.

42E

Ou plutôt l'acide et l'alkali[15] se conjoignent pour ne plus se séparer, à cause de la conformité de leur Nature.

43E

Tant que l'Alkali[16] domine, dure le regne[17] de sécheresse ; mais l'acide prédominant à son tour fait regner[18] l'humidité.

44E

La prédomination de l'acide entraîne la dissolution du corps, et améne[19] la fermentation.

---

14. dessiccation.
15. l'alcali.
16. l'Alcali.
17. règne.
18. régner.
19. amène.

39TH

To moisten and desiccate make up the natural day.

40TH

Each humectation, or act of moistening, is called cohabation, and each desiccation, distillation.

41ST

With each imbibition, the central fire retains the spiritual portion of the Secret Fire; the phlegm dissipates completely.

42ND

Or rather the acid and the alkali are conjoined, never to be separated, due to the conformity of their Nature.

43RD

As long as the Alkali dominates, the reign of dryness endures; but when, in its turn, acid predominates, it causes moisture to reign.

44TH

The predominance of acid induces the dissolution of the body, and brings on fermentation.

45E

Cette fermentation n'est qu'un combat entre l'acide et l'alkali[20] pendant lequel ils se tuent l'un l'autre.

46E

L'acide a pourtant surmonté le fixe puisqu'il l'a amené à dissolution; mais le fixe a aussi vaincu l'esprit volatil qui demeure sans action.

47E

De l'acide et de l'alkali[22] réunis se forme une nature androgine[23] ou hermaphrodite.

48E

La fermentation achevée, la Putréfaction vient à la suite, et met le Sceau au premier travail.

49E

Il y eut 50 Néréides ou déesses des humidités 50 filles de Danaüs[23] qui épousèrent les 50 fils d'Ægyptus.

50E

Il faut 50 ablutions de l'esprit sur la terre, ou 50 mariages de l'acide et de l'alkali,[24] du ciel avec la terre, pour obtenir la dissolution.

---

20. l'acali.
21. l'alcali.
22. androgyne.
23. Danaüs.
24. l'alcali.

45TH

This fermentation is but a combat between acid and alkali during which they destroy each other.

46TH

However, the acid has overcome the fixed, since it has brought about its dissolution; but the fixed has also conquered the volatile spirit, which remains without efficacy.

47TH

United together, the acid and alkali form an androgynous or hermaphroditic nature.

48TH

The fermentation being completed, Putrefaction follows, and places the Seal on the first work.

49TH

There were fifty Nereids or sea goddesses,[2] fifty daughters of Danaus who married the fifty sons of Ægyptus.

50TH

There must be fifty ablutions[3] of the spirit upon the earth, or fifty marriages of acid and alkali, of heaven with earth, to obtain the dissolution.

---

2. Literally 'goddesses of moisture', 'goddesses of humidity'.
3. That is, 'washings'.

51E

L'alkali[25] faisant fonction de femelle, surmonte 49 fois son mâle qui est l'esprit ; mais, à la 50.eme les forces venant à lui manquer, il demeure conjoint.

52E

On cesse les ablutions aussitôt que la fermentation se presente.[26] On compare ce feu au bain Marie.[27]

53E

La chaleur augmentant dans la putréfaction est comparée à celle du fumier.

54E

Ce n'est que dans la putréfaction que la conjonction est opérée. Le[28] principes renfermés dans une seule substance ne peuvent plus être séparés, et c'est ce qu'on appele[29] Sceau Hermétique.

55E

Du charbon qui est noir on fait de la cendre grise, et de cette cendre on tire un sel par continuation du feu.

56E

Le corps noirci[30] par putréfaction devient gris et est comparé aux cendres, ensuite blanc, et est le vrai sel de nature ou le salpêtre des sages, c'est à dire[31] le Sel de leur pierre.

---

25. l'alcali.
26. présente.
27. bain-Marie.
28. Les.
29. appelle.
30. noircit.
31. c'est-à-dire.

51ST

Taking the female function, the alkali overcomes its male, which is the spirit, forty-nine times. But with the fiftieth, the forces have failed him, and he remains a spouse.

52ND

The ablutions are ceased as soon as fermentation occurs. We compare this fire to the Bain-Marie.

53RD

The increase of heat during putrefaction can be compared to that of manure.

54TH

It is only in the putrefaction that the conjunction is effected. The principles contained in a single substance can no longer be separated, and this is what we call the Hermetic Seal.

55TH

From the coal which is black we make grey ash, and from this ash we extract a salt by maintaining the fire.

56TH

The body darkens by putrefaction, becomes grey, and is compared to ashes; it then becomes white and is the true salt of nature or the saltpetre of the wise, i.e. the Salt of their stone.

57E

Les sages comparent encore leur matière au savon, parce que indépendament[32] de ses propriétés particulières elle est comme le savon composée d'un alkali[33] auquel la graisse du Soufre est jointe.

58E

Dans la cendre, disent les sages, est renfermé le Diadême[34] de notre jeune roi ; dans la terre restante, après l'extraction du sel, est le soufre.

59E

Le soufre se manifeste dans cette terre par sa coction avec notre esprit ou feu Secret.

60E

Les philosophes appelent,[35] feu externe, l'administration de l'esprit au corps, de l'acide à l'alkali[36] ou l'excitation produite entre le sel et l'humide.

61E

Geber définit la sublimation l'élévation par le feu d'une chose sèche avec adhérence au vaisseau, pour exprimer la putréfaction et exaltation de la substance, le feu, la chose sèche, et le vase étant ensemble une même chose.

---

32. indépendamment.
33. alcali.
34. Diadème.
35. appellent.
36. l'alcali.

57TH

Once again the sages compare their matter to soap, because regardless of its particular properties, it is like soap because it is made from an alkali to which the fat of Sulphur is joined.

58TH

In ash, the sages say, is contained the Diadem of our young king. In the earth that remains after the extraction of the salt, is the sulphur.

59TH

Sulphur manifests in this earth through its coction with our spirit, or Secret Fire.

60TH

The philosophers call 'external fire' the administering of spirit to the body, of acid to the alkali, or the excitation produced between salt and wetness.

61ST

Geber[4] defines sublimation as the elevation by fire of something dry which adheres to the vessel, in order to express that the putrefaction and exaltation of the substance, the fire, the dry thing, and the vase are all one and the same thing.

---

4. JĀBIR IBN HAYYĀN, Arab alchemist, c. 721–c. 825. The Latin GEBER, which is referred to here, is mostly known for his *Summa Perfectionis* (William NEWMAN has recently identified this as a work of a certain Paul of Taranto, a 13th-century Franciscan alchemist).

62e

Le sel des sages a besoin d'être exalté pour devenir leur mercure. Ils comptent neuf sublimations.

63e

Les sublimations se font comme le premier travail, par l'administration du feu externe.

64e

Le mercure doit être fait par le Mercure, c'est à dire[37] que le feu doit être de même substance que le corps soumis au travail.

65e

Pour que cela soit ainsi, il faut dissoudre dans l'esprit une partie du sel pour faire les Imbibitions.

66e

A cet effet, on fait, à chaque sublimation deux parts de son Sel, l'une demeure sèche, et on dissout l'autre pour imbiber.

67e

Il se fait ainsi une nouvelle dissolution, fermentation et putréfaction d'autant plus prompte que le sel est plus élevé en dignité.

68e

Ces sublimations que Philalette[38] nomme ses aigles, ne peuvent outrepasser le nombre de neuf.

---

37. c'est-à-dire.
38. Philalèthe.

62ND

The salt of the sages needs to be exalted to become their mercury. They count nine sublimations.

63RD

Sublimations are done as in the first work, by administering the external fire.

64TH

Mercury must be made by Mercury, i.e., the fire must be of the same substance as the body submitted to the work.

65TH

For this to happen, it is necessary to dissolve a part of the salt in the Spirit to make the Imbibitions.

66TH

To this end, with each sublimation, we separate its Salt into two parts. One remains dry, and we dissolve the other in order to imbibe.

67TH

It thus forms a new dissolution, fermentation, and putrefaction, and all the more swiftly, for the salt is of a higher dignity.

68TH

These sublimations, which Philalethes calls his eagles, cannot exceed the number of nine.

69e

A chaque sublimation du Sel de nature ou mercure, il se sépare toujours, au moyen de la dissolution, un peu de terre qu'il faut réunir à la première.

70e

Ce sont toutes ces terres réunies que l'on met avec notre esprit, pour avoir le soufre.

71e

Dans cette coction, il n'y a ni dissolution, ni fermentation, ni putréfaction à attendre, le corps ne fait que rougir de plus en plus et arrive à une couleur brune qui est la dernière.

72e

Pour avoir cette Teinture couleur de sang qui est l'or solaire, ou le vinaigre très aigre, ou esprit de vin de R. Lulle &c. il faut verser dessus la terre rouge, le mercure pphique[39] à la hauteur de deux à trois doigts; alors elle se sépare doucement et surnage le mercure comme une Quintessence.

73e

Lorsqu'on dissout avec l'esprit astral, le sel qui est le mercure, il faut mettre la dissolution dans un lieu frais, le mercure se rassemble alors sur la superficie de l'esprit sous forme de crême,[40] mais c'est un sel, ou une eau sèche qui, bien que liquide ne mouille pas les mains.

---

39. philosophique.
40. crême.

69TH

With each sublimation of the Salt of nature, or mercury, a little earth is always separated through the process of dissolution, and it must be combined with the first.

70TH

It is all these earths combined that are placed with our spirit, in order to obtain the sulphur.

71ST

In this coction, neither dissolution, fermentation, nor putrefaction is to be expected. The body does nothing but redden increasingly until it changes into a brown colour, which is the last.

72ND

In order for this Tincture to become the colour of blood, which is the solar gold, or very sharp vinegar, or the spirit of wine of R. Lulle etc., it is necessary to pour the philosophical Mercury over the red earth to a height of two to three fingers. It then separates gently and floats on the mercury like a Quintessence.

73RD

When one dissolves the salt, which is the mercury, with the astral Spirit, the dissolution must be placed somewhere cool. The Mercury then gathers on the surface of the spirit in the form of a cream, but it is a salt, or a dry water which, although liquid, does not wet the hands.

74E

Il reste dans l'esprit deux sortes de sels autres que le mercure: savoir un sel nitreux et un sel fixe.

75E

En faisant subir à ces sels le travail des aigles, et les travaillant l'un par l'autre, ils arrivent tous deux à une forme parfaite mercurielle.

76E

Il y a deux voies pour avoir le soufre; la voie humide, et la voie sêche.[41]

77E

La voie humide est celle que je viens d'enseigner, c'est la plus longue, mais la plus noble, à cause des difficultés vaincues.

78E

La voie sèche, telle que Flamel et B. Trévisan l'on suivie, méne[42] au même but; quoique particulière.

79E

Elle consiste à séparer la Teinture de l'or commun avec le mercure du $7.^{me}$ aigle.

80E

On gagne ainsi sur le tems[43] deux sublimations du mercure et la coction entière de la terre des aigles.

---

41. sèche.
42. mène.
43. temps.

74TH

There remains in the spirit two kinds of salts other than mercury: namely a nitrous salt and a fixed salt.

75TH

While subjecting these salts to the work of the eagles, and working them one by one, they both arrive at a perfect mercurial form.

76TH

There are two ways to obtain the sulphur; the wet way, and the dry way.

77TH

The wet way is that which I have just taught. It is the longest, but also the noblest, due to the difficulties that are overcome.

78TH

The dry way, as followed by Flamel and B. Trevisan, leads to the same goal; it is, however, specific.

79TH

It involves separating the Tincture from common gold using the Mercury of the seventh eagle.

80TH

This saves time on two sublimations of mercury as well as the entire coction of the earth of the eagles.

81e

Quand on a procédé par la voie humide, il faut rejeter la terre qui reste après l'extraction de la Teinture. C'est une terre damnée et domageable.[44]

82e

Quelle que soit la voie que vous suiviez, il faut procèder[45] au mariage du soufre et du Mercure.

83e

Ce sont de Nouveaux Cieux et une Nouvelle Terre que vous allez marier ensemble, et qui produiront une nouvelle Jerusalem avec un roi très puissant.

84e

Prenez une partie de votre Soufre ou Teinture, laissez la[46] sécher et il s'en formera une terre très douce et agréable au toucher, d'un rouge brun.

85e

Faites avec le mercure vos imbibitions, comme à la première opération, en suivant le poids de Nature.

86e

Après 40 Imbibitions qui sont réputées 40 jours, le corps se dissoudra, fermentera et pourrira.

---

44. dommageable.
45. procéder.
46. laissez-la.

81ST

When one proceeds by the wet way, one must reject the earth which remains after the Tincture has been extracted. It is an accursed and deleterious earth.

82ND

Whatever way that you follow, it is necessary to proceed to the marriage of Sulphur and Mercury.

83RD

It is a New Heaven and a New Earth that you will wed together, which will produce a new Jerusalem with a very powerful king.

84TH

Take part of your Sulphur or Tincture, let it dry, and it will form a reddish-brown earth that is very soft and pleasant to the touch.

85TH

Make your imbibitions with mercury, as in the first operation, according to the weight of Nature.

86TH

After forty Imbibitions, which is considered to be forty days, the body will dissolve, ferment, and putrefy.

87E

Ce sont ces deux Teintures, l'une rouge l'autre blanche que *Le Petit-Paysan* nomme ses deux fleurs, et que d'autres ont appelées Grande et petite Lunaire.

88E

Il faut amener cette teinture rouge à la blancheur par imbibitions avec le mercure.

89E

Ces imbibitions doivent être faites de manière que la terre demeure ferme, quoique mouillée.

90E

La Science demeure dans les principes; mais l'art consiste à savoir dissoudre et pourrir.

91E

Celui là[47] est passé maître qui a atteint le degré de putréfaction, car quoique ce soit le plus bas de l'Œuvre, il est réputé le plus élevé à cause des difficultés qu'il présente pour y arriver.

92E

Le principal pas est fait pour arriver à la blancheur qui est une médecine souveraine contre toute sorte de maux.

---

47. Celui-là.

87TH

It is these two Tinctures, one red and the other white, that *Le Petit-Paysan*[5] calls his two flowers, and which others have called the Great Moon and the Lesser Moon.

88TH

It is necessary to bring this red tincture to whiteness by imbibitions with Mercury.

89TH

These imbibitions must be made in such a way that the earth remains firm, yet moist.

90TH

Science remains based on principles; but art consists in knowing how to dissolve and putrefy.

91ST

He who has reached the degree of putrefaction is a Master, because although this is the lowest Work, it is deemed to be the highest due to the difficulties involved in order to attain it.

92ND

The main step has been taken to bring about the whiteness, which is a sovereign medicine against all kinds of ailments.

---

5. 'The Little Farmer', a French translation from the German text, *Der grosse und kleine Bauer* (The Great and Little Farmer), a treatise attributed to Johann GRASSHOFF. See our comments to *Hermetic Recreations*, p. 79 n. 46, in the present volume.

93E

Ce n'est pas qu'il ne reste quelques difficultés à vaincre, mais elles ne sont pas insurmontables.

94E

On n'arrive pas de suite à la blancheur, il faut auparavant dissoudre et noircir.

95E

Il faut que ce soit une dissolution radicale, et que le corps soit réduit en ses plus menues parties, bien qu'il ne soit pas semblable à de l'eau fluviale ou semblable.

96E

C'est à tort que quelques pphes[48] ont parlé de noircir le blanc, car quoique la blancheur sorte de la noirceur, c'est néanmoins le rouge que l'on a blanchi et le même par conséquent que l'on a noircit.[49]

97E

Au surplus cette noirceur est un voile ténébreux qui couvre la blancheur aussi bien que la rougeur.

98E

On appele[50] la dissolution le Sceau de mercure, le bain marie,[51] le bain du Roi. Quant à la putréfaction dont la noirceur est le symbole, c'est le fumier de bouc ou de cheval, et le Sceau de Saturne.

---

48. philosophes.
49. noirci.
50. appelle.
51. bain-Marie.

93RD

Although some difficulties remain to be overcome, they are not insurmountable.

94TH

We do not reach the whiteness straight away; it must first dissolve and blacken.

95TH

It must be a radical dissolution, and the body must be reduced to its tiniest parts, but it should not flow like river water or similar.

96TH

Certain philosophers have been wrong to speak of blackening the white, for although whiteness comes out of blackness, it is nevertheless the red which is whitened and consequently the same which is blackened.

97TH

Furthermore this blackness is a dark veil, which covers the whiteness as well as the redness.

98TH

We call dissolution the Seal of Mercury, the bain-marie, or the bath of the King. As for the putrefaction, whose symbol is blackness, it is the manure of a goat or horse, and the Seal of Saturn.

99E

La dissolution est prise par les uns pour la première matière des sages, et la putréfaction par les autres, eu égard à la réunion essentielle et inséparable des deux substances.

100E

Quoi qu'il en soit, la dissolution est proprement le cahos[52] des sages, dans lequel le Ciel et la Terre sont renfermés, et la putréfaction est leur principee[53] matière.

101E

Ce n'est qu'au bout de 40 imbibitions que le corps se dissout, fermente et pourrit.

102E

On appele[54] Tête de Corbeau, Saturne ou Plomb des pphes[55] cette première noirceur.

103E

Comme au 1.er travail on cesse d'administrer le feu externe, lorsque la dissolution est entière.

104E

La matière se conduit par son propre feu jusqu'au cercle de la blancheur qui est la lune des pphes,[56] Diane, Latone,[57] ou le Laton blanchi.

---

52. chaos.
53. principale.
54. appelle.
55. philosophes.
56. philosophes.
57. Laiton.

99TH

Dissolution is taken by some for the first matter of the wise, and putrefaction by others, given the essential and inseparable reunion of the two substances.

100TH

In any event, dissolution is, properly speaking, the chaos of the sages in which Heaven and Earth are contained, and putrefaction is their principal matter.

101ST

It is only at the end of forty imbibitions that the body dissolves, ferments, and putrefies.

102ND

One calls this first blackness the 'Head of the Raven', or 'Saturn', or the 'Lead of the philosophers'.

103RD

As with the first work, one stops administering the external fire when the dissolution is complete.

104TH

The matter is conducted by its own fire to the 'circle of whiteness', which is the moon of the philosophers, Diana, Latona, or whitened Brass (Latten).[6]

---

6. See commentary to *Hermetic Recreations*, p. 69 n. 38, in the present volume.

105E

La blancheur commence par un cercle capilaire[58] qui s'étend de jour en jour jusqu'au centre; mais avant d'arriver à la blancheur, la matière passe du noir au gris qui est la couleur intermédiaire et qu'on nomme le feu de cendre, et le Sceau de Jupiter.

106E

Le passage du gris au blanc est marqué par l'apparition de plusieurs couleurs, parmi lesquelles domine la verte: ce qui a fait donner à la blancheur le nom de Lion vert.

107E

Les sages nomment ces couleurs Iris, ou Queue de Paon.

108E

On compare ce travail jusqu'à la blancheur, au feu de Reverbére.[59]

109E

La blancheur, que nous avons dit être le regne[60] de la Lune, n'est qu'une demi-génération. Les sages l'appelent[61] terre feuillée pour deux raisons principales.

110E

1° C'est que quand on la regarde de prés,[62] elle ressemble à des feuilles de Talc brillant.

---

58. capillaire.
59. Réverbère.
60. règne.
61. l'appellent.
62. près.

105TH

The whiteness begins with a capillary circle that grows, day by day, towards the centre. But before arriving at the whiteness, the matter passes from black to grey, which is the intermediate colour, and which is called the 'fire of ash', and the 'Seal of Jupiter'.

106TH

The passage from grey to white is marked by the appearance of several colours, among which green predominates; because of this, the name 'green Lion' has been given to the whiteness.

107TH

The wise call these colours Iris, or the the Tail of the Peacock.

108TH

This work, up until the whiteness, is compared to the fire of Reverberation.

109TH

Whiteness, which we have said to be the reign of the Moon, is only a half generation. The wise call it 'foliated earth'[7] for two principal reasons.

110TH

The first is that when one inspects it closely, it resembles sheets of brilliant Talc.

---

7. That is, earth of 'leaflike layers'.

111E

2° C'est que la putréfaction où elle vient de passer est le symbole de l'hiver pendant lequel la terre est couverte de feuilles dont une nouvelle terre se forme au printems,[63] laquelle terre est appelée terre des feuilles.

112E

La matière ne pouvant aller plus loin par son propre feu, il faut recommencer le feu externe.

113E

Pour se préparer d'avance à la multiplication, il faut séparer en deux la matière.

114E

On en met une part de côté, et on conduit l'autre à la rougeur, en continuant le travail.

115E

On reprend donc ici le travail des imbitions avec le mercure, observant les poids de Nature.

116E

Il faut comme la premiere[64] fois que la terre demeure entiére[65] au fond du vaisseau.

---

63. printemps.
64. première.
65. entière.

111TH

The other is because the putrefaction, which it has just passed through, is the symbol of winter, during which the earth is covered with leaves from which a new earth is formed in spring. This earth is called the 'earth of leaves' ('foliated earth').

112TH

The matter cannot go any further by its own fire. It is necessary to begin the external fire again.

113TH

To prepare in advance for multiplication, it is necessary to separate the matter into two.

114TH

One part is set aside, and the other is brought to redness by continuing the work.

115TH

We thus resume here the work of the imbibitions with mercury, observing the weights of Nature.

116TH

Like the first time, the earth must remain completely intact at the bottom of the vessel.

117e

La matière perd peu à peu sa blancheur et arrive à une couleur verte que l'on compare au Vitriol, et que l'on appèle[66] le Sceau de Vénus.

118e

Par la continuité du feu, elle acquiert une couleur jaune saffranée[67] qui est le Sceau de Mars.

119e

La matière ne pouvant par le même degré de feu acquerir[68] une plus grande rougeur, il faut l'augmenter.

120e

On augmente le feu en imbibant le corps avec le mercure Rouge mis en réserve.

121e

On continue de cette manière jusqu'à ce que la matière ait acquis un Rouge brun.

122e

Avant d'arriver à cette rougeur foncée, elle passe à une belle couleur de pourpre.

---

66. appelle.
67. safranée.
68. acquérir.

117TH

The matter gradually loses its whiteness and attains a green colour that we compare to Vitriol, and which we call the 'Seal of Venus'.

118TH

By the persistence of the fire, it acquires a saffron yellow colour, which is the 'Seal of Mars'.

119TH

The matter can no longer acquire a greater redness by the same degree of fire; it must be increased.

120TH

We increase the fire by soaking the body with the Red Mercury kept in reserve.

121ST

We continue in this manner until the matter acquires a Reddish-Brown colour.

122ND

Before arriving at this deep redness, it passes through a beautiful purple colour.

123E

La matière arrivée au rouge brun très foncé, est le vrai Or ou fluide des sages, leur soleil, leur médecine universelle.

124E

Sauf les multiplications, il n'y a plus de difficultés à vaincre.

125E

On possède deux médecines; l'une blanche et l'autre rouge pour guérir toute maladie.

126E

Ces deux médecines ne sont pas seulement utiles aux hommes, mais aux végétaux et aux minéraux.

127E

Un arbre presque mort arrosé d'eau dans laquelle sera dissout un seul grain pesant de cette médecine, reprendra vie, fleurira, et fructifiera.

128E

On fait avec cette médecine une infinité de merveilles au dessus[69] du pouvoir naturel.

---

69. au-dessus.

123RD

The matter, having reached a very deep reddish brown, is the true, fluid Gold of the wise, their Sun, their Universal Medicine.

124TH

Except for the multiplications, there are no more difficulties to overcome.

125TH

We possess two medicines; one white and the other red, to cure any disease.

126TH

These two medicines are not only useful to men, but also to plants and minerals.

127TH

A tree which is nearly dead, sprinkled with water in which only a single grain measure of this medicine is dissolved, will revive, blossom, and bear fruit.

128TH

With this medicine we make an infinite number of wonders beyond natural power.

129E

Si vous projetez un grain de la médecine blanche sur dix de bon argent, le tout sera médecine dont un grain en transmuera 100 en métaux imparfaits, en argent meilleur que celui des mines.

130E

Un grain de médecine rouge projeté sur de bon or en fusion, faira[70] de l'or dans la même proportion.

131E

Pour faire des perles plus grosses et plus belles que les naturelles, on n'a besoin que d'en dissoudre avec le mercure et de les mouler ensuite.

132E

On augmente de même manière le poids et la beauté du Diamant et des pierres précieuses.

133E

On fait des Rubis artificiels, bien plus éclatans[71] que les naturels, par addition de teinture Rouge.

134E

Mais il n'y a que dieu seul qui puisse rappeler les corps de la mort à la vie.

---

70. fera.
71. éclatants.

129TH

If you project a grain of white medicine onto ten of fine silver, all of them will become a medicine; and from this, one granule will transmute 100 granules of imperfect metal into silver superior to that of the mines.

130TH

A grain of red medicine projected onto fine molten gold will make gold in the same proportion.

131ST

To make pearls larger and more beautiful than natural ones, we only need to dissolve some of them with Mercury, and then mould them.

132ND

In the same manner, we can increase the weight and the beauty of Diamonds and precious stones.

133RD

By adding some Red Tincture, we can make artificial Rubies far more brilliant than natural ones.

134TH

But it is God alone who can call the bodies of the dead back to life.

135E

La teinture Rouge est le septieme[72] et le dernier Sceau d'Hermès qui appartient au Soleil.

136E

On procède à la multiplication avec des parens[73] d'un même sang.

137E

On appèle[74] parent d'un même sang les taintures[75] blanche et rouge d'une même opération.

138E

Le mercure qui n'a pas été accouplé avec la tainture[76] Rouge, n'est pas propre à multiplier.

139E

Les médecines blanche et Rouge du 1.er degré sont parents d'un même sang, et peuvent multiplier.

140E

C'est dans cette intention qu'on sépare les médecines en deux, dans les Cercles de la blancheur et de la rougeur.

---

72. septième.
73. parents.
74. appelle.
75. teintures.
76. teinture.

135TH

The Red Tincture is the seventh and final Seal of Hermes which belongs to the Sun.

136TH

We proceed to the multiplication with the parents of the same blood.

137TH

We call the parents of the same blood the white and red tinctures of one and the same operation.

138TH

Mercury, which has not been coupled with the Red Tincture, is not proper for multiplying.

139TH

The White and Red medicines of the first degree are parents of the same blood, and can multiply.

140TH

It is with this intention that we separate the medicines into two, in the Circles of whiteness and redness.

141E

On procède à la première multiplication en prenant une part de teinture rouge qu'on dissout avec la blanche mise en réserve.

142E

Il faut auparavant dissoudre la blanche avec le mercure pour procèder[77] aux Imbibitions.

143E

On recommence alors le premier travail avec les mêmes conditions et observant le poids de nature.

144E

Le pur séparé de l'impur abrège chaque fois de moitié le tems[78] de l'opération.

145E

La projection de cette seconde médecine se fait sur 100 d'argent ou d'or, comme ferment, et ensuite sur mille des métaux imparfaits.

146E

Le poids et la vertu de la médecine augmentant de dix à chaque multiplication, une once, de la neuvieme,[79] transmuera[80] un million en très pur métal d'or ou d'argent.

---

77. procéder.
78. temps.
79. neuvième.
80. transmutera.

141ST

We proceed to the first multiplication by taking a part of the red tincture, which we then dissolve with the white tincture that was set aside.

142ND

It is necessary to dissolve the white with Mercury in order to proceed with the imbibitions.

143RD

We then start the first work again with the same conditions, observing the weights of Nature.

144TH

Each time, the pure separates from the impure in half the time of the previous operation.

145TH

This second medicine is projected onto a hundred parts of silver or gold, as a ferment, and then onto a thousand parts of imperfect metal.

146TH

The weight and the virtue of the medicine increases by ten with each multiplication. One ounce of the ninth (multiplication) will transmute a million (ounces) into highly pure metallic silver or gold.

147E

La vertu de cette médecine est si grande qu'elle peut en un instant changer de face toute la nature sublunaire.

148E

C'est pour que les méchans[81] n'en approchent pas que les sages la tiennent si cachée.

149E

Passé la neuviéme[82] multiplication la médecine ne peut plus être contenue; elle flue à travers le verre, comme l'huile à travers le papier.

150E

L'Œuvre entier s'achève en 150 jours, excepté les multiplications qui peuvent conduire à deux cents.

fin

---

81. méchants.
82. neuvième.

147TH

The virtue of this medicine is so great that it can instantly change the whole face of sublunary nature.

148TH

It is in order to keep the wicked away that the wise keep it so hidden.

149TH

After the ninth multiplication the medicine can no longer be contained. It seeps through glass like oil through paper.

150TH

The entire Work is completed in 150 days, except for the multiplications, which can extend it to two hundred.

end

Bibliography

Manuscript

Les Récréations hermétiques, Ms. 362, Muséum d'Histoire Naturelle, Paris, pp. 1039–1054.
Scholies, Ms. 362, Muséum d'Histoire Naturelle, Paris, pp. 1055–1064.

Publications

Artephius, *Artefii Arabis Philosophi, Liber Secretus. Nec non Saturni Trismegisti, siue Fratris Heliæ de Assisio Libellus*, Francfurti, Iennisium, 1685.
Azaïs, Pierre Hyacinthe, *Des Compensations dans les destinées humaines*, Paris, Garnery, 1809.
Berthelot, Marcellin, and C. E. Ruelle, *Collection des anciens alchimistes grecs*, 3 vols., Paris, Georges Steinheil, 1888–1889.
Böke, Christer, 'Efterord' to Fulcanelli, *Katedralernas mysterium: en esoterisk tolkning av de hermetiska symbolerna i det stora alkemiska verket*, Malmö, Vertigo, 2013.
———., and John Koopmans, 'Fulcanelli's Most Likely Identity—Part I', *Alchemy Journal*, vol. 7, no. 2, 2006.
———., and John Koopmans, 'Fulcanelli's Most Likely Identity—Part II', *Alchemy Journal*, vol. 7, no. 3, 2007.
Boyle, Robert, *The Sceptical Chymist, or Chymico-Physical*

Doubts & Paradoxes, Touching the Spagyrist's Principles, Commonly call'd Hypostatical, as they are wont to be Propos'd and Defended by the Generality of Alchemists, London, J. Crooke, 1661.

«CENT CINQUANTE SCHOLIES», in *La Tourbe des Philosophes*, issue 14–16, Paris, La Table d'Émeraude, 1981.

CHEAK, Aaron, 'The Alchemy of Desire: The Metaphysics of Eros in René Schwaller de Lubicz (A Study of *Adam l'homme rouge*)', in H. T. HAKL, ed., *Octagon, Volume 2: The Quest for Wholeness, Mirrored in a Library Dedicated to Religious Studies, Philosophy and Esotericism in Particular*, Gaggenau, Scientia Nova, 2016.

COHAUSEN, Johann Henrich, *Hermippus Redivivus: Or the Sage's Triumph Over Old Age and the Grave*, London, J. Nourse, 1748.

_____. *Hermippus redivivus, ou le triomphe du sage, sur la vieillesse et le tombeau ; contenant une méthode pour prolonger la vie et le vigueur de l'homme*, translated by M. De La Place, Bruxelles, Maradan, 1789.

CYLIANI, *Hermès dévoilé dédié à la postérité*, Paris, Imprimerie de Félix Locquin, 1832; Paris, Chacornac, 1915; B. Husson, ed., Paris, L'Omnium Littéraire, 1964

D'ESPAGNET, Jean, *Enchiridion physicæ restitutæ, in, quo verus naturæ concentus exponitur, plurimique antiquæ philosophiæ errores per canones & certas demonstrationes dilucidè aperiuntur; Tractatus alter inscriptus, Arcanum Hermeticæ philosophiæ opus, in quo occulta Naturæ & Artis circa lapidis philosophorum materiam & operandi modum canonicè & ordinatè fiunt manifesta*, Paris, Nicolaum Buon, 1623.

DICKINSON, Edmund, *Epistola Edmundi Dickinson MD. & medici regii ad Theodorum Mundanum philosophum adeptum De quintessentia philosophorum et De vera physiologia, una cum quæstionibus aliquot de secreta materia physica : his accedunt Mundani responsa*, Londini, E Theatro Sheldoniano, 1686.

DUJOLS, Pierre, *Hypotypose du Mutus Liber*, Paris, Editions Nourry, 1914.

FLAMEL, Nicholas, *Le Sommaire philosophique*, in Jean de MEUN(G), ed., *De la transformation métallique*, Paris, Guillaume Guillard, 1561.

———. *Le Livre des figures hiéroglyphiques*, in *Trois traitez de la philosophie naturelle*, Paris, Guillaume Guillard et Amaury Warancore, 1561.

FULCANELLI, *Le Mystère des cathédrales, et l'interprétation ésotérique des symboles hermétiques du Grand-Œuvre*. Paris, Jean Schmidt, 1930; Éditions Alcor, 2013.

———. *Les Demeures philosophales et le symbolisme hermétique dans ses rapports avec l'art sacré et l'ésotérisme du Grand-Œuvre*, Paris, Jean Schmidt, 1930; Éditions Alcor, 2014.

———. *The Dwellings of the Philosophers*, translated by Brigitte Donvez and Lionel Perrin, Boulder, Archive Press, 1999.

———. *The Mystery of the Cathedrals*, translated by Mary Sworder, Las Vegas, Brotherhood of Life, 2003.

———. *Katedralernas mysterium : en esoterisk tolkning av de hermetiska symbolerna i det stora alkemiska verket*, translated by Kjell Lekeby, Malmö, Vertigo, 2013.

GAGNON, Claude, « Découverte de l'identité de l'auteur réel du 'Livre des figures hiéroglyphiques' », *Anagrom : Sorcellerie, alchimie, astrologie*, Maisonneuve et Larose, no 7-8, 1976, p. 106.

———. *Nicolas Flamel sous investigation, suivi de l'édition annotée du Livre des Figures Hiéroglyphiques*, Québec, Éditions Le Loup de gouttière, 1994.

GODWIN, Joscelyn, *The Theosophical Enlightenment*, Albany, State University of New York Press, 1994.

GRASSHOFF, Johann, *Aperta Arca arcani artificiosissimi. Das ist: Der allergrösten vnd künstlichsten Geheimnüssen der Natur eröffneter vnd offstehender Kasten: Darinnen von der waren Materia, vnnd deren, deß einzigen Subjecti*

Vniuersalis magni, vnfehlbaren Erkantnuß, auß welchem allein das höchste Werck, nemlich der lapis Philosophicus, welches allen Tincturen in der gantzen Chymia vorgehet, entspringet, auch auß dessen Geist alle dinge der gantzen Welt vniuersaliter herfliessen, klärlich gehandelt wirdt. Beneben der rechten vnd warhafftigen Physica Natvrali Rotvnda, durch eine Visionem Chymicam Cabalisticam gantz verständtlich beschrieben ; Allen Gottesfürchtigen ... mitgetheilet vnd an tag geben, Franckfurt, Bringer, 1617.

HAKL, H. T., ed., *Octagon, Volume 2: The Quest for Wholeness, Mirrored in a Library Dedicated to Religious Studies, Philosophy and Esotericism in Particular*, Gaggenau, Scientia Nova, 2016.

HUSSON, Bernard, ed., *Deux traités alchimiques du XIXe siècle*, Paris, L'Omnium Littéraire, 1964.

_____. *Anthologie de l'Alchimie*, Paris: P. Belfond, 1971.

I.C.H., ed., *Des Hermes Trismegists wahrer alter Naturweg. Oder: Geheimniß wie die große Universaltinctur ohne Gläser, auf Menschen und Metalle zu bereiten*, Leipzig, A. F. Böhme, 1782.

KIM, Mi Gyung, *Affinity, That Elusive Dream: A Genealogy of the Chemical Revolution*, Cambridge, MIT Press, 2003.

KINGSLEY, Peter, 'From Pythagoras to the *Turba Philosophorum*: Egypt and Pythagorean Tradition', *Journal of the Warburg & Courtauld Institutes*, 1994, vol. 57, pp. 1-13.

KLOSSOWSKI DE ROLA, Stanislas, *Alchemy: The Secret Art*, London, Thames & Hudson, 1973.

_____. *Alchimie ; florilège de l'art secret. Augmenté de La fontaine des amoureux de science*, Paris, Seuil, 1974.

_____. *The Golden Game: Alchemical Engravings of the Seventeenth Century*, London, Thames & Hudson, 1988.

_____. *Le Jeu d'or : figures hiéroglyphiques et emblèmes hermétiques dans la littérature alchimique du XVIIe siècle*, Paris, Thames & Hudson, 1997.

LAVOISIER, Antoine, *Traité élémentaire de chimie, présenté*

dans un ordre nouveau, et d'après des découvertes modernes, Paris: Cuchet, 1789.

LENGLET DUFRESNOY, Nicolas, *Histoire de la Philosophie Hermetique*, 3 volumes, Paris, Coustelier, 1742.

LIMOJON DE SAINT-DIDIER, Alexandre-Toussaint de, *Le Triomphe hermétique ou la Pierre philosophale victorieuse*, Amsterdam, Henry Wetstein, 1690/1699.

LUCAS, Paul, *Voyage du Sieur Paul Lucas fait par ordre du Roi dans la Grece*, Nicolas Simart, Amsterdam, 1714.

LUX OBNUBILATA *suapte natura refulgens*, Venice, 1666; *La Lumière sortant par soi-même des Ténèbres*, with commentary by Bruno de Lansac, 1687; ed., Bernard Roger, Paris, E. P. Denoël, 1971.

MEUN(G), Jean de, ed., *De la transformation métallique. trois anciens tractez en rithme françoise*, Paris, Guillaume Guillard et Amaury Warancore, 1561.

NEWMAN, William R., *The Summa Perfectionis of Pseudo-Geber: A Critical Edition, Translation, and Study*, Leiden, Brill, 1991.

PARACELSUS, Philippus Aureolus Theophrastus Bombastus von Hohenheim, *Sämtliche Werke*, edited by Karl Sudhoff, 14 volumes, R. Oldenbourg, Munich & Berlin, 1929–1933.

PASQUIER, Gilles, *L'Entrée du labyrinthe*, Paris, Éditions Dervy, 1992.

PETRUS BONUS, *Margarita Preciosa*, Venice, Giovanni Lacinio, 1546; also in *Introductio in divinam chemicæ artem, integra magistri Boni Lombardi Ferrariensis physici*, Michael Toxites, ed., Basel, Pietro Perna, 1572.

POTTS, Malcom, 'Etymology of the Genus Name Nostoc (Cyanobacteria)', *International Journal of Systematic Bacteriology*, vol. 47, no. 2, 1997, p. 584.

ROOS, Anna Marie, 'Johann Cohausen (d. 1750), Volatile Salts, and Theories of Longevity in a German Satire', *Medical History*, vol. 51, no. 2, 2007, pp. 181–200.

SCHWALLER DE LUBICZ, R. A., *Adam l'homme rouge, ou les*

éléments d'une gnose pour le mariage parfait, St. Moritz, Officina Montalia, 1927.

SENDIVOGIUS, Michael, *alias* THE COSMOPOLITAN, *Tractatus de sulphure altero naturæ principium*, Colloniæ, Apud Ioannem Chrithium, 1616.

RUSKA, Julius, *Turba Philosophorum: Ein Beitrag zur Geschichte der Alchemie*, Berlin, Julius Springer, 1931.

TREVISAN, Bernard. *Trevisanus de Chymico miraculo, quod lapidem philosophiæ appellant*, Basel, Pietro Perna, 1583.

TROIS TRAITEZ DE LA PHILOSOPHIE NATURELLE *non encore imprimez : scavoir Le secret livre du tres-ancien philosophe Artephius, traitant de l'art occulte et transmutation metallique, Latin François : plus Les figures hierogliphiques de Nicolas Flamel ainsi qu'il les a mises en la quatriesme arche qu'il a bastie au Cimitiere des Innocens à Paris, entrant par la grande porte de la ruë S. Denys, & prenant la main droite, avec l'explication d'icelles par iceluy Flamel ; ensemble, Le vray livre du docte Synesius Abbé grec, tiré de la bibliotheque de l'empereur sur le mesme subiect*, Paris, Guillaume Marette, 1612.

TURBA PHILOSOPHORUM, *Auriferae artis, quam chemiam vocant, antiquissimi authores, sive Turba philosophorum*, Basel, Pietro Perna, 1572.

_____. *Or the Assembly of the Sages; Called also the Book of Truth in the Art and the Third Pythagorical Synod*, translated by Arthur Edward Waite, London, George Redway, 1896.

VALENTINUS, Basilius. *Ein kurtz summarischer Tractat, von dem grossen Stein der Uralten*, Eisleben, Hörnig, Bartholomaeus, 1599.

_____. *Occulta philosophia*, Johann Bringern, 1613.

_____. *Von den natürlichen und übernatürlichen Dingen. Auch von der ersten Tinctur, Wurzel und Geiste der Metallen und Mineralien wie dieselben empfangen ausgekocht, geboren, verändert und vermehret werden*, in *Chymische Schrifften*, Hamburg, 1677.

———. *Chymische Schrifften*, Hamburg, 1677.
———. *Les Dovze clefs de philosophie*, Elise Vveyerstraten, Amsterdam, 1678.

ZADKIEL THE ALCHEMIST, a.k.a. John PALMER, *The Familiar Astrologer*, William Bennett, London, 1831.

ABOUT THE CONTRIBUTORS

Christer Böke and John Koopmans have collaborated regularly during the past fifteen years in the intensive historical research and interpretation of ancient and modern alchemical texts. Together, they published the controversial article, 'Fulcanelli's Likely Identity' in the «Alchemy Journal» (2006–2007). They plan to prepare and release future translations and transcriptions of important, rare, and obscure alchemical texts that will greatly assist the interested researcher. Relevant feedback can be sent to: saturniaduplex@gmail.com

CHRISTER BÖKE (MA, History and Theology) has been researching the history of alchemy for the past two decades. His major interest lies in reconstructing alchemical theories from an experimental perspective. Christer wrote his Masters thesis (Lund, 2002) on the 'Paracelsus dispute' (1707–1708) between the swedish chymists, Urban Hjärne and Magnus Gabriel von Block, triggered by the infamous Paykull affair—a topic he further covered during 'On the Fringes of Alchemy', an alchemical workshop held with different scholars in Budapest 2010. Christer participated in the Swedish translation and commentary of Fulcanelli's *Le Mystère des Cathédrales* (*Katedralernas mysterium*, 2013), and acted as a peer-reviewer on Carl-Michael Edenborg's alchemical fiction drama, *Alkemistens dotter* (2014).

JOHN KOOPMANS (BA, Hons, Geography) is a retired professional demographer. Over the years, he has collaborated in, or contributed to, a number of books concerning biblical history, and the historical study of Grail lore. When he's not intensely studying Alchemy, or transcribing numerous old alchemical texts (many in original handwriting), he continues to maintain a very long, active interest in various studies, including recent and ancient history (civilisations, archæology, religions, sciences, philosophies, esoterica, and mythologies).

PRINCE STANISLAS KLOSSOWSKI DE ROLA (Baron de Watteville) inspired a reevaluation of the alchemical tradition with his two books, *Alchemy: The Secret Art* (1973) and *The Golden Game* (1988), and their French elaborations, *Alchimie ; florilège de l'art secret* (1974), and *Le Jeu d'or* (1997). Son of Balthus, the renowned Polish-French aristocrat and artist, Stanislas is a man of unique extraction and influence. A close personal friend to Eugène Canseliet—the direct disciple of the legendary adept, Fulcanelli—he more famously befriended The Beatles and The Rolling Stones in the 1960s, and gained popular attention for his dandyist æsthetic. He later lived for many years in Sri Lanka, and was personally acquainted with Lama Anagarika Govinda, an important authority on the deeper links between eastern and western wisdom traditions. He presently resides in the Castle of Montecalvello, Italy.

AARON CHEAK, PHD, is a scholar of comparative religion, philosophy, and esotericism. In 2011, he received his doctorate in religious studies for his work on French Hermetic philosopher, René Schwaller de Lubicz, and from 2013–2015 he served as president of the International Jean Gebser Society. Founder and director of Rubedo Press, he has appeared in both academic and esoteric publications, including *Alchemical Traditions* (2013), *Clavis*

(2014), *Diaphany* (2015), *Lux in Tenebris* (2016), and *The Leaf of Immortality* (2017). He currently lives on the west coast of New Zealand, where he maintains an active interest in tea, wine, poetry, typography, and alchemy.

www.ingramcontent.com/pod-product-compliance
Lightning Source LLC
Chambersburg PA
CBHW021953290426
44108CB00012B/1052